JAMSHID GHARAJEDAGHI

THE SYSTEMS INQUIRY SERIES

Systems inquiry is grounded in a philosophical base of a systems view of the world. It has formulated theoretical postulates, conceptual images and paradigms, and developed strategies and tools of systems methodology. Systems inquiry is both conclusion oriented (knowledge production) and decision oriented (knowledge utilization). It uses both analytic and synthetic modes of thinking and it enables us to understand and work with ever increasing complexities that surround us and which we are part of.

The Systems Inquiry Series aims to encompass all three domains of systems science: systems philosophy, systems theory and systems technology. Contributions introduced in the Series may focus on any one or combinations of these domains or develop and explain relationships among domains and thus portray the systemic nature of systems inquiry.

Five kinds of presentations are considered in the Series: (1) original work by an author or authors, (2) edited compendium organized around a common theme, (3) edited proceedings of symposia or colloquy, (4) translations from original works, and (5) out of print works of special significance.

Appearing in high quality paperback format, books in the Series will be moderately priced in order to make them accessible to the various publics who have an interest in or are involved in the systems movement.

SERIES EDITORS

BELA H. BANATHY and GEORGE KLIR

TOWARD A
SYSTEMS THEORY
OF ORGANIZATION

TOWARD A SYSTEMS THEORY OF ORGANIZATION

JAMSHID GHARAJEDAGHI

THE WHARTON SCHOOL
University of Pennsylvania

The Systems Inquiry Series
PUBLISHED BY INTERSYSTEMS PUBLICATIONS

Copyright ©1985 by Intersystems Publications
All Rights Reserved
Published in the United States of America by Intersystems Publications
Printed in U.S.A.

This publication may not be reproduced, stored in a retrieval system, or transmitted in whole or in part, in any form or by any means, electronic, mechanical, photocopying, recording, or otherwise, without the prior written permission of INTERSYSTEMS PUBLICATIONS (Seaside, California 93955, U.S.A.).

ISBN 0-914105-35-3

CONTENTS

PREFACE   v

FOREWORD by RUSSELL L. ACKOFF   vii

1. MECHANISMS, ORGANISMS AND SOCIAL SYSTEMS   1
   with R. L. ACKOFF

2. CULTURE AS STRUCTURE OF SOCIAL SYSTEMS   17

3. SOCIAL DYNAMICS: DICHOTOMY OR DIALECTIC   29

4. ON THE NATURE OF DEVELOPMENT   49

5. OBSTRUCTIONS TO DEVELOPMENT   63

6. ORGANIZATIONAL IMPLICATIONS OF SYSTEMS THINKING   75

7. PERFORMANCE CRITERIA AS A MEANS OF SOCIAL INTEGRATION   97
   with ALI GERANMAYEH

BIBLIOGRAPHY   113

To RUSSELL L. ACKOFF,
of whose teaching this is
a small reflection.

Acceptable ideas are competent no more.

Competent ideas are not yet acceptable.

This is the dilemma of our time.

Stafford Beer

PREFACE

The papers in this book were designed to stand alone. Nevertheless, together they are a whole, which is more than sum of its parts.

Despite unconventionality and the theoretical tone, all of the ideas presented here have emerged out of practice — twenty years of real life experimentation with organizations in different cultures.

Fascinated by questions of why and how, I have been searching to find some answers to why social systems behave the way they do and how can one get a handle on them.

Frustrated and dissatisfied with the answers provided by prevailing and conventional models of organizations, I came to the conclusion that models based on mechanistic or organismic analogies cannot provide the necessary insight. Social systems ought to be understood in their own right.

At this stage of confusion I rediscovered Russell Ackoff and his key concept of purposeful systems. This exciting conception opened up a whole new vision. But before I could begin to understand its real implications I had to do a lot of unlearning. In addition, K. Boulding's concept of "image" combined with Walter Buckley's "complex adaptive systems" provided me with the insight to distinguish between "information-bonded" and "energy-bonded" systems. Subsequently, the synthesis of *purposefulness* with information *bondedness* became the core of the social systems model presented in this work.

The first paper written with R. Ackoff gives an overview of the three basic models of organizations.

For further elaboration of the systems model, I have followed my bias that a holistic conception of a social system requires an understanding of its structure, process, and functions. The next three chapters are devoted to an exploration of these three aspects of a social system.

After dealing briefly with the system of problems (messes) in the fifth chapter, the next two chapters deal with the "how" question by introducing an organizational design and explaining the practical implications of the assumptions introduced in the previous chapters.

I wish to express my indebtedness to my dear colleagues in the Social Systems Sciences Department — Tom Cowan, West Churchman, Hasan Ozbekahn, Aron Katsenelinboigen, Iraj Zandi, and Jean-Marc Choukroun — for their support, encouragement, and critiques, which I found invaluable. I have taken most of their advice, but I am solely responsible for the outcome. Associates in Busch Center, expecially Alex Ackoff, Roberta Snow, Ali Geranmayeh, Johny Pourdehnad, and Omid Nodoushani, have given me so generously of their time that I consider them to be the actual co-producers of this work. I am also particularly grateful to my administrative colleagues in the Busch Center, Pat Brandt and Joan Lonetti, for providing me with ideal conditions for work. Over time I have learned more from my friends, clients, and students than I have taught them. For the debts that I am not able to acknowledge I must beg to be forgiven.

This acknowledgement, however, would not be complete if I neglected to mention the influences and support of my old friends during my exciting years of tenure in the Industrial Management Institute. Special thanks are due to Reza Niazmand, Reza Ghotbi, Akbar Etemad, Nader Hakimi and Bijan Khoram.

Finally, it is not repetitious to express my gratitude to Russell L. Ackoff who has made all this possible.

Jamshid Gharajedaghi
October, 1984

FOREWORD

by
RUSSELL L. ACKOFF

There is nothing that an author who has tried to produce new ideas values more than having another take those ideas and develop them further. Jamshid Gharajedaghi has done just this to my work. But he has done a great deal more; he has made significant additions of his own.

The tradition out of which his work has come and that from which mine has arisen are very different, but these two traditions intersected a number of years ago and have merged to give his work a freshness and originality that I envy. It may be helpful to the reader to share some of the history from which Jamshid's and my joint efforts have emerged.

I began graduate work in the philosophy of sciences at the University of Pennsylvania in 1941 where I came under the influence of the "grand old man" of the department, the eminent philosopher E. A. Singer, Jr. Because of the informality of the department he created I began to collaborate with two younger members of the faculty both of whom were former students of Singer, Thomas A. Cowan and C. West Churchman.

Three aspects of Singer's philosophy had a particularly strong influence on me. First, that the practice of philosophy, its application, was necessary for the development of philosophy itself. Second, that effective work on "real" problems required an interdisciplinary approach. Third, that the social area needed more work than any of the other domains of science and that this was the most difficult.

We developed a concept of a research group that would enable us to practice philosophy in the social domain by dealing with real problems. The organization we designed was called "The Institute of Experimental Method." With the participation of a number of other graduate students in philosophy and a few other members of the faculty we started this Institute on a completely informal basis.

In June of 1946 I accepted an appointment to the Philosophy Department of (then) Wayne University in Detroit. I did so because the dean of the college had shown enthusiasm for the idea of establishing an Institute of Applied Philosophy and offered to support an effort to create it.

In the following year Churchman also accepted a full-time appointment in philosophy. Meanwhile, Cowan had immigrated to the Law School of Wayne from Nebraska to which he had gone when he left Penn in 1946.

The other two members of the philosophy department of Wayne viewed our efforts to establish an Institute of Applied Philosophy as prostitution of this ancient pursuit. A "fight" broke out over this issue, one that involved a large part of the faculty, administration, and student body at Wayne. My position in that department became untenable.

In the Spring of 1951 Churchman and I accepted appointments to (then) Case Institute of Technology in Cleveland because Case was committed to establishing an activity in Operations Research and Churchman and I had come to believe we could probably work better under this name than under the cloak of academic philosophy. By the end of 1952 we had formal approval, but not without faculty opposition, of the first doctoral program in Operations Research. From then on the Group and the program grew rapidly and flourished. Case became a mecca to which pilgrimage of operations researchers from around the world came.

In 1958, Churchman, for personal reasons, migrated to the University of California at Berkeley where he established a similar activity. Academic Operations Research activities began to proliferate and flourish, many of them modeled on those at Case.

In June of 1964 the research group and academic program moved to Penn bringing with it most of the faculty, students, and research projects.

Our activities flourished in the very supportive environment that Penn and Wharton provided. The wide variety of faculty members that we were able to involve in our activities significantly enhanced our capabilities.

By the mid 1960s I had become uncomfortable with the direction, or rather, the lack of direction, of professional Operations Research. I had five major complaints. First, it had become addicted to its mathematical tools and had lost sight of the problems of management. As a result it was looking for problems to which to apply its tools rather than looking for tools that were suitable for solving the changing problems of management.

Second, it failed to take into account the fact that problems are abstractions extracted from reality by analysis. Reality consists of systems of problems, problems that are strongly interactive, messes. I believed that we had to develop ways of dealing with these systems of problems as wholes.

FOREWORD

Third, Operations Research had become a discipline and had lost its commitment to interdisciplinarity. Most of it was being carried out by professionals who had been trained in the subject, its mathematical techniques. There was little interaction with the other sciences, professions, and humanities.

Finally, Operations Research was ignoring the developments in systems thinking — the methodology, concepts, and theories being developed by systems thinkers.

For these reasons, five of us on the OR faculty designed a new program which we wanted to provide as an option to students entering the program. In addition to myself, there was Eric Trist, Hasan Ozbekhan, Thomas Saaty, and James Emshoff. We were able to initiate a new experimental program and administrative entity in The Wharton School called the Social Systems Sciences. It came to be known as "S Cubed." This program along with its research arm, the Busch Center, now hosts the largest doctoral program in the School.

The graduate and research programs are directed at producing professionals who were capable of planning for, doing research on, and designing social systems, systems in which people play the major role. It is dedicated to the development and use of theories of social systems and professional practice, and the practice of such theories. It is also committed to the development of methodology and conceptual systems which enable us to design and manage social systems more effectively.

In 1968 I made my first trip to Iran on a mission for the U.N. I met Jamshid during that visit. He was then employed by IBM. On one of my subsequent visits I found that he had assumed the direction of the Industrial Management Institute and had integrated the research and academic principles of S^3 with its own Program developed locally.

We started a personal and institutional collaboration. He sent a number of his Staff to us for graduate work and we engaged in several joint projects. We tried to entice him to Penn as a visiting Professor but he was unwilling to leave his remarkable Institute. I could not blame him. In his position I would have acted as he did. Unfortunately for him, but fortunately for us, the revolution in Iran changed all that. That upheavel virtually destroyed his Institute and his opportunities for carrying out his work. He left Iran with the help of our invitation and immediately joined us. Shortly after, I was able to transfer the Direction of the Busch Center to him.

His joining us was a major event in my life. An investigator into a serious and complex subject welcomes a *convergence* of a broad stream of ideas, experience and hard work of a distinctively different cultural origin.

This book is a record of collaboration between the system of systems thought stemming originally from the works of Edgar A. Singer, T. Cowan, C. West Churchman, and myself, working primarily in the cultural milieu of the Western world and the author of this book working for many years in the apparently quite dissimilar situation of an ancient Eastern culture.

An apparent miracle happened. What was originally thought of as a fundamentally disparate source of alien views on the nature of systems organization turned easily and naturally into a joint effort. The fundamental nature of systems organization was at once perceived to be a unity in diversity.

When Professor Gharajedaghi joined the Social Systems Sciences Department of the Wharton School and assumed the direction of its research arm, the Busch Center, he began a two-pronged activity of research into the nature of systems organization and applied research and application. In a series of his writings on systems theory it became evident quite early that the two streams of thought were not only basically compatible but also had the happy effect of enriching each other. The evidence of this fortunate coalescent of a different cultural reapproachment is the present work.

Jamshid is not only an invaluable friend and colleague, but is also a constant source of inspiration. Therefore, I was delighted by the invitation to open this book which enables me to invite you to share in the inspiration he has provided me.

Russell L. Ackoff
The Wharton School
University of Pennsylvania

1

MECHANISMS, ORGANISMS AND

SOCIAL SYSTEMS*

with RUSSELL L. ACKOFF

We live in an age of accelerating change, increasing uncertainty, and growing complexity. People are being pushed and pulled by forces that previously did not seem to be part of their environment. They respond by acquiring more information and knowledge, *but not understanding*. In fact, many futurists characterize the post-industrial era we are entering as an age of information or knowledge (e.g., Naisbitt, 1982), not as an age of understanding.

Information is *descriptive*; it is contained in answers to questions that begin with such words as *what*, *which*, *who*, *how many*, *when*, and *where*. Knowledge is *instructive*; it is conveyed by answers to *how-to* questions. Understanding is *explanatory*; it is transmitted by answers to *why* questions. To understand a system is to be able to explain its properties and behavior, and to reveal why it is what it is and why it behaves the way it does.

Information, knowledge, and understanding form a hierarchy. Information presupposes neither knowledge nor understanding. Knowledge presupposes information, and understanding presupposes both. One can survive without understanding, but not thrive. Without understanding one cannot control causes; only treat effects, suppress symptoms. With understanding one can *design and create the future*.

To think about anything requires an image or concept of it, a model. To think about a thing as complex as a social system most people use a model of something similar, simpler, and more familiar. Traditionally, two types of model have been used in efforts to acquire information, knowledge, and understanding of social systems: *mechanistic* and *organismic*. But in a world of accelerating changes, increasing uncertainty, and growing complexity it is becoming apparent that these are inadequate as guides to deci-

---

*This paper originally appeared in Strategic Management Journal, Vol. 5 1984, (John Wiley & Sons, Ltd.) and is reproduced here by permission of the editor.

sion and action. Alienation, hopelessness, frustration, insecurity, corruption, tyranny, and social unrest are only a few of the many symptoms of deeply rooted malfunctioning of societies and their institutions. Commonly prescribed remedies are increasingly ineffective and often make things worse. The growing number of social crises and dilemmas that we face should be clear evidence that something is fundamentally wrong with the way we think about social systems.

Here we describe and try to explain the deficiencies of the two traditional ways of thinking about social systems. The principal deficiency is the limited yield of understanding. We then develop a third type of model, one we believe does not suffer from these inadequacies, a *social system* model.

THE MECHANISTIC MODEL

Mechanistic models of the world conceptualize it as a machine that works with a regularity dictated by its internal structure and the causal laws of nature. The world, like a hermitcally sealed clock, is taken to be made up of purposeless and passive parts that operate predictably. Any deviation from regularity is reacted to with changes that restore it; the system is believed to tend in the long run toward a static equilibrium.

This type of model is based on two assumptions: the world can be *completely understood* and such understanding can be obtained by *analysis*. Analysis is a three-step thought process. First, it takes apart that which it seeks to undertand. Then it attempts to explain the behavior of the parts taken separately. Finally, it tries to aggregate understanding of the parts into an explanation of the whole.

Since understanding something mechanistically requires understanding its parts, the parts also have to be taken apart. This process stops only when indivisible parts, elements are reached. These, when understood, are believed to make understanding everything else possible. This doctrine, called *reductionism*, is responsible for the prominence in science of such irreducibles as atoms, chemical elements, cells, basic needs, instincts, direct observations, and phonemes.

Once the elements are understood, their explanations have to be aggregated into an understanding of the whole. This requires establishing a relationship between the parts. The relationship that is assumed to be sufficient to explain all actions and interactions of the parts is *cause-effect*. One thing is taken to be the cause of another if it is both necessary and sufficient for the other.

The exclusive commitment to cause-effect has three important consequences. First, because identification of causes provides *complete* explanations of their effects, the environment is not required to explain anything. This environment-free concept of explanation is reflected in such natural laws as that of *freely* falling bodies, which applies only in the absence of an environment. It is also reflected in the predisposition to conduct research in laboratories, places from which the environment can be excluded.

Second, causes themselves require explanation. This is done by treating them as effects and finding their causes which also must be explained. Is there no end to this regression? Given the mechanist's assumption that the world is completely comprehensible, the answer must be "yes"; there has to be a *first cause*. This was generally taken to be God and naturally, He was taken to be the *Creator*. God alone is uncaused and, therefore, not subject to explanation. He must be accepted on faith.

Third, because of the assumed comprehensibility of the world everything other than God has to be assumed to be the effect of some cause, and therefore to be *determined* by that cause. Such determinism leaves no room for choice, hence purpose, in the natural world.

The effects of applying mechanistic models to social systems are manifested by adherents to the so-called classical or traditional school of management. The way they organize work is a direct consequence of analytical thinking. They begin by reducing work to elementary tasks, tasks so simple that they can be performed by one person alone. The simplicity of these tasks facilitates their mechanization. Only those tasks that are too expensive or complex to be mechanized are assigned to people. Work is reduced to machine-like behavior and workers are treated like replaceable machine parts.

Adherence by parts to rules and regulations is made an end-in-itself either by rewarding compliance or punishing noncompliance. By this means, human responses to stimuli are made to approximate mindless physical reactions.

Control and coordination are also analyzed and reduced to tasks requiring and minimal amount of power and judgement at each organizational level. Judgement is further reduced by establishing policies that offer virtually no choice except to determine which policy applies to which situation.

Like the universe, believed to have been created by God to do His work, organizations are taken to be instruments of their owners with no purposes of their own. Corporations, for example, are considered to be instruments for producing profit for their owners.

Mechanistically modeled organizations are structured hierarchically and are centrally controlled by a completely autonomous authority. Such an authority can affect any part of the system without being affected by any of them. This separates the ultimate authority from the system making that authority an external controller.

All members of the system other than the one with ultimate authority are deprived of all information except that required to do their jobs. Instructions from above are not explained or justified. Blind adherence is expected: "Theirs is not to reason why; theirs is but to do or die." People are kept apart as much as possible. Their interactions are minimized to depersonalize their relations. Even non-work related interactions are discouraged.

The operations of an ideal machined do not vary. Therefore, as long as its input does not vary, its output won't. For this reason controllers of mechanistically modeled social system focus on inputs rather than outputs. For example, they assume that control of costs is equivalent to control of outputs. The quantity of output is assumed to be determined by the quantity of input. The system is thought of much as a vending machine.

A mechanistically conceived social system is inflexible. Therefore, it can operate effectively only if its environment is static or has little effect on it; that is, where it can operate as a *closed* system. However, a rapidly changing environment requires continuous adaptation and learning by organizations if they are to remain effective. Adaptation and learning require a readiness, willingness, and ability to change, but these are precisely what mechanistically managed and structured organizations lack.

The larger social system of which every organization is a part, and other organizations that are part of that containing system, no longer permit an organization to ignore its effects on and the effects on it of its environment. Recognition of this interaction poses a severe dilemma to mechanistically conceptualized organizations. They find it difficult or impossible to be responsive to environmental changes. As a result, their effectiveness suffers. Increasing ineffectiveness leads to reinforcement of their regidity, closer adherence to rules and regulations. The result is a vicious circle in which such organizations become more and more dysfunctional (March and Simon, 1958).

THE ORGANISMIC MODEL

A social system conceptualized as an organism has a purpose of its own: *survival* for which *growth* is taken to be essential. Contraction is believed to be synonymous with deterioration and decay, with eventual death. Such a system is taken to be dependent on this environment for essential inputs (resources). Therefore, if that environment is believed to be changing, to survive the system must be capable of learning adaptation.

Growth is necessary but not sufficient for survival. Nothing can preclude eventual death, but continuity through reproduction can keep death from being terminal. Therefore, such systems try to reproduce themselves either by spawning new organizations (e.g., establishing colonies) or by acquiring old one (imperialism).

In an organismically conceptualized corporation, profit, like oxygen in the case of an organism, is taken to be necessary for survival but not the reason for it. Profit is taken as a means; growth as an end. With survival as the ultimate end, planning becomes prediction of environmental changes and preparation for them.

Preoccupation with growth creates certain problems: first, the fact that things can grow only at the expense of other systems and their environment and, second, that exponential growth, the best kind, cannot be sustained forever. Moreover, there seems to be an optimal size for each type of organization beyond which an increase in size leads to a decline in efficiency and effectiveness (Boulding, 1970).

Because changes in their environments are considered to be inevitable and relevant, organismically conceptualized social systems seek a dynamic rather than a static equilibrium. They operate homeostatically, adjusting the behavior of their parts to maintain the properties of the whole within acceptable limits. Their parts are thought of as organs, each with a function the performance of which contributes to the survival and growth of the whole. Individuals are regarded as cells whose function it is to serve the organs and the organism of which they are part. Organs and cells are more difficult to replace than machines or machine parts.

The executive function is thought of as the brain of the system (Beer, 1981). It is linked to the parts of the system by a communication network through which it receives information from a variety of sensing organs (e.g., diplomatic corps, intelligence services, and marketing departments), and issues directives that activate and deactivate the parts of the system.

An organismically conceived social system is organized hierarchically but, since thinking and sensing are separated, control is not as completely centralized as it is in a mechanistically conecived system. Different parts have some degrees of self-control, but they do not control the functions they are intended to carry out. Some parts can interact directly with and, in some cases, react directly to environmental changes without intervention of "the brain." As well as formal structure, organismic social systems have an informal one that is maintained by direct communication between parts. There is also more two-way communication between and within different levels of the hierarchy in an organismic social system than in a mechanistic one. Moreover, conformity and obedience of the parts is not taken to be as essential as long as they perform well. They are managed by control of outputs rather than inputs.

Organismic organizations try to exercise control by specifying desired outputs, leaving the selection of means to the parts (Management by Objectives). The environment and organizational performance (outputs) are kept under surveillance to determine whether they are behaving as expected. If not, corrective action is taken. Thus, management engages in feedback control. This facilitates both learning and adaptation.

Although organismically conceived systems are capable of self-control, they can be influenced by other systems and, in some cases, can be controlled by them by the application of force. Force may be required to make such an organization act against its will.

A horse, unlike an automobile, cannot be driven into a wall without compulsion. External control of an organism is easiest where there is submission or consent, both of which are matters of choice.

To treat an organization or other type of social system as an organism is to fail to recognize the ways in which these differ significantly. In contrast to an organism, which cannot change its structure more than a limited amount and still survive, a social system has almost complete control over its structure (Buckley, 1967). In addition, the relationship that exists between an organism and its cells and organs is very different from that between an organization and its parts. One's heart cannot decide for itself that it does not want to work or wants to work for someone else. The parts of a social system have purposes of their own and display choice. Therefore, an effective social system requires agreement among its parts and between its parts and the whole. It requires consenses; an organism does not.

An organismically conceived system devotes itself to making the best of a future that it believes to be largely out of its control, but is predictable. The fact is that in an environment characterized by accelerating change increasing uncertainty, and growing complexity, the possibility of accurate and reliable forecasts diminishes at an increasing rate. In such an environment the only hope for a social system lies in its ability to bring more and more of its future under its own control. To take such an approach requires a model of a social system different from the two we have reviewed.

THE SOCIAL SYSTEM MODEL

A system is a whole that cannot be divided into independent parts; the behavior of the parts and their effects on the whole depends on the behavior of other parts. Therefore, the essential properties of a system are lost when it is taken apart; for example, a disassembled automobile does not transport and a disassembled person does not live. Furthermore, the parts themselves lose their essential properties when they are separated from the whole; for example, a detached steering wheel does not steer and a detached eye does not see.

Recall that analysis takes a system apart and then tries to explain the behavior of the parts taken separately. It is for this reason that it cannot yield understanding of a system, only knowledge of how it works. Put another way: it can reveal a system's structure but not its functions.

Because the parts of a system are interdependent, it can be shown that even if each part is independently made to perform as efficiently as possible, the system as a whole will not perform as effectively as possible. For example, all-star athletic teams are not known to be the best teams and may not be able to beat an average team in the league.

The performance of a system is not the sum of the independent performances of its parts. It is the product of their interactions. Therefore, effective management of a system requires management of the interactions of its parts, not their independent actions. Moreover, since a social system interacts with its environment, management of this interaction is also required for it to function effectively.

It follows, then, that to understand a system as a whole a new non-analytical approach is needed. Contrary to a widely held belief a multidisciplinary approach does not fill this need for reasons revealed in the classical story of the blind men trying to identify an elephant, each by touching a different part of its body. The interesting point in this story is not the predicament of the blind men but the ability of the story-teller to see the whole. A different version of the same story found in old Persian literature (molana Jalalledin Molalvi) is about a group of men who encounter a strange animal in complete darkness. Their efforts to identify it, each by touching a different part, prove fruitless until someone arrives with a lamp. Light enables us to synthesize separate observations into a meaningful whole. This light must be provided by a way of thinking other than analysis.

To understand a system its structure, processes, and functions have to be examined. A system's structure is the way its work is divided among its parts and their efforts coordinated, that is, the relationships between its parts. Structure can be understood only if observed in the functioning of a system. Therefore, analysis, which revelas only the structure of a system, not its functioning, cannot provide understanding, only knowledge. *Synthetic* thinking is required to complement analysis.

In the first step of analysis the thing to be explained is taken apart; in synthetic thinking it is taken as a part of a larger system. In the second step of analysis the contained parts are explained; in synthetic thinking the containing system is explained. In the third step of analysis an effort is made to aggregate understanding of the parts into an understanding of the whole; in synthetic thinking understanding of the containing whole is disaggregated to explain the parts by revealing their role or function in that whole. In the social systems model synthetic thinking and analysis are taken to be complementary; neither can replace the other. Both are necessary to understand a system.

To understand the functioning of a social system the cause-effect relationship is inadequate. The *producer-product* relationship (E.A. Singer, Jr., 1959) is required. A producer is only ncecessary, not sufficient, for its product. Therefore, a producer never completely explains its product; it does *not* determine its product. This makes it possible to treat choice, hence purpose, as an objectively observable property of system behavior. Moreover, since a producer is not sufficient for its product, reference to its environment is required to explain its product. Therefore, explanation become *environment-full* rather than environment-free as it is in the mechanistic model.

As a consequence, knowledge of the processes by which parts of a system choose to respond to their environment is essential to understanding the dynamics of a system which changes its structure. Process, rather than initial conditions, is responsible for future states; similar conditions (initial states) can lead to dissimilar outcomes (end states).

In a structuralist conception of a system a specific structure ($\underline{S}$) causes a particular function ($\underline{F}$), and different structures cause different functions (Figure 1). Therefore, to understand a system, knowledge of its structure (the nature of its components and their relationships) is taken to be all that is required. This structuralist point of view is based on a deterministic conception of reality.

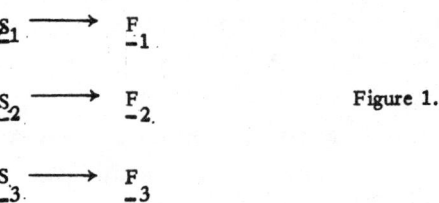

Figure 1.

In a social systemic conception of a social system (1) different structures can yield the same function, and (2) the same structure can yield different functions (Figure 2). This makes it necessary to understand the relationship between structure

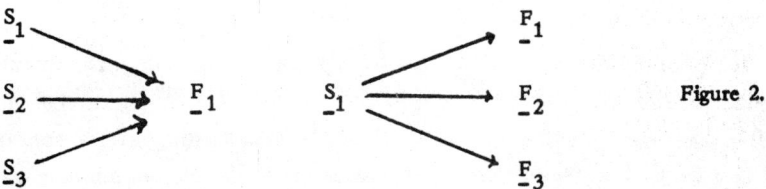

Figure 2.

and function keeping in mind that function is not a substitute for structure any more than form is a substitute for content. They are complements.

The time-telling function can be produced by such different structures as sun dials, and sand, water, mechanical, or electronic clocks. The transportation function can be produced by cars, ships, trains, and automobiles. An educational system with a specific structure in a particular environment, in addition to its explicit educational function, also provides daytime child care. Clocks often have a decorative function as well as time-telling. Therefore, it is important to note that a redesign of a system to increase its efficiency with respect to one of its functions may, and often does, decrease its efficiency with regard to its other functions, and therefore its overall effectiveness.

Now, what is a function? According to Ackoff and Emery (1972) a *functional class* is:

> ... a set of structurally different individuals, systems or events, each of which is either a potential or actual producer of members (objects or events) of a specified class ($\underline{Y}$) of any type.
>
> The function of such a class is ($\underline{Y}$)-production, and each member of the class can be said to have $\underline{Y}$-production as its function. If an individual displays a type of structural behavior that is a member of a functional class of behavior displayed by other entities, then that individual or system can be said to have an *extrinsic function*. It is called extrinsic because the function is not one of its own but one it has by virtue of its membership in a class. The property that forms such a class is not structural but a common property of production ...
>
> A person who can telephone a store, write to it, visit it, or get another to visit it displays a set of actions (events) that constitute a functional class defined by, say, acquiring a new shirt. In this case the individual displays an *intrinsic function* because its function can be attributed to it on the basis of *its behavior alone* (p. 26).

Using the difference between cause-effect and producer-product, Ackoff and Emery also distinguish between three types of system behavior: *reactions*, *responses*, and *actions*.

A *reaction* of a system is a system event for which another event that occurs to the same system or its environment is *sufficient*. Thus a reaction is an event that is (deterministically) caused by another event.

A *response* of a system is a system event for which another event that occurs to the same system or its environment is *necessary but not sufficient*. Thus a response is an event of which the system itself is a *co-producer*. A person's turning on a light when it gets dark is a response to darkness, but the light's going on when the switch is turned is a reaction.

An *act* of a system is a system event for the occurrence of which no change in the system's environment is either necessary or sufficient. Acts, therefore, are *self-determined* events, autonomous behavior.

Systems all of whose behavior are reactive, responsive, or active can be called reactive, responsive, or active, respectively. However, most systems display some combination of these types of behavior. Nevertheless, it should be noted that mechanically conceptualized systems are modeled as predominantly reactive; organismically conceptualized systems as predominantly responsive; and socially conceptualized systems as predominantly active. (See Table 1.)

Reactive, responsive, and active systems are also correlated with *state-maintaining*, *goal-seeking*, and *purposeful* systems.

A *state-maintaining* system is one that *reacts* to changes so as to maintain its state under different environmental conditions. Such a system can react (not respond) because

TABLE 1. BEHAVIORAL CLASSIFICATION OF SYSTEM

TYPE OF SYSTEM	BEHAVIOR (Means, structure)	OUTCOME (Ends, functions)
Passive (Tools)	*Fixed* One structure in all environments.	*Fixed* One function in all environments.
Reactive (Self-maintaining system)	*Variable but Determined* Once structure in any one environment, different structures in different environments.	*Variable but Determined* One function in any one environment, different functions in different environments.
Responsive (Goal-seeking)	*Variable and Chosen* Different structures in the same environment.	*Variable but Determined* One function in any one environment, different functions in different environments.
Active (Purposeful) system)	*Variable and Chosen* Different structures in the same environment.	*Variable and Chosen* Different functions in the same environment.

what it does is completely determined by the change in its environment, given the structure of the system. Nevertheless, it can be said to have the intrinsic function of maintaining the state it maintains because it can produce this state in a different way under different conditions.

For example, many heating systems are state-maintaining. An internal controller turns it on when the room temperature goes below a desired level, and turns it off when the temperature goes above this level. The state that is maintained is the room temperature. Such a system is able to adapt to changes but is not capable of learning because it cannot choose its behavior. It cannot improve with experience.

A *goal-seeking* system is one that can *respond* differently to one or more different events in one or more different environments, and that can respond differently to a particular event in an unchanging environment until it produces a particular outcome (state). Production of this state is its goal. Such a system has a choice of behavior, hence is responsive rather than reactive. Response is voluntary; reaction is not.

For example, lower-level animals can try to get a food in different ways in different environments or even in the same environment at different times.

If a goal-seeking system has memory it can learn to pursue its goal more efficiently over time.

A *purposeful* system is one that can produce the same outcome in different ways in the same environment and can produce different outcomes in the same and different environments. It can change its ends under constant conditions. This ability to change ends under constant conditions is what exemplifies *free will*. Such systems can not only learn and adapt; they can *create*. Human beings are examples of such systems.

Purposeful systems have all the capabilities of goal-seeking and state-maintaining systems, and goal-seeking systems have the capabilities of state-maintaining systems. The converse is not true. Therefore, they form a hierarchy.

Social systems are purposeful systems. Moreover, their parts are purposeful systems, and they are part of larger social, hence purposeful, systems. To understand a social system, then, one must not only know what the ends of the parts, system, and containing systems are, but how these affect their interactions. Managing a social system not only requires dealing with ends that may be in conflict at the different levels, but dealing with conflicting ends at any or all of the levels.

Recall that a mechanistically conceived system is not attributed with a purpose of its own even though it may serve a purpose of an external controller. Organismic systems have a purpose of their own: survival, for which growth is taken to be necessary. Then what is the appropriate purpose of a system conceptualized as a social system? *Development.*

DEVELOPMENT

Growth and development are not the same thing and are not even necessarily associated. Either can take place with or without the other. A cemetery can grow without developing; a person may continue to develop long after he or she has stopped growing. This, of course, is obvious. What is not so obvious is that many of the problems associated with development derive from the assumption that economic growth is necessary if not sufficient for development and that limits to growth limits development.

Growth, strictly speaking, is an increase in size or number. Its principal but not exclusive domain of relevance is biological, as in the growth of plants and animals. Social systems are said to grow when the economic value of their product or the amount of income they generate increases. Growth, however, is often used metaphorically or figuratively, and when it is, it frequently and incorrectly is taken literally. For example, when we speak of a person "growing up" we refer to increased maturity. Maturity has no size or number. It would be nonsensical to speak of a growing culture or quality of life because size and number are not relevant to them. A social system, like an individual, can grow by increasing in size or, unlike an individual, in number without developing. It can also develop without growing.

Growth occurs naturally, without choice, in most biological systems. In human beings and social systems, however, choices can deter or accelerate growth as, for example, in choice of diet or investment portfolio. If someone has a compulsion to grow, to increase in size, we are likely to consider this person to be a pathological case. For example, medical science increasingly views obesity as the product of a psychiatric disorder. However, if an organization has a compulsion to grow we generally consider it to be natural, even laudable. Why? Because we assume that physical or economic growth is necessary, if not sufficient, for development. This, as we will try to show, is not the case. Nevertheless, if limits to growth threatened an organization's survival, one could understand its preoccupation with such limits, but not even the authors of *The Limits to Growth* (Meadows, et al, 1972) claim that this is so. Unfortunately, many have incorrectly drawn the inference from this book that retarded growth implies retarded development. Such an error, we believe, is based on a misconception of the nature of development.

Development of a person, contrary to what many believe, is not a condition or a state defined by what or how much that person has. For example, if the products and service available in the most affluent nation were suddenly showered on an aborigine, he would not thereby be developed. Development has less to do with how much a person has than with *how much he or she can do with what he does have*. For this reason, Robinson Crusoe and the Swiss Family Robinson are better paradigms of development than J.P. Morgan and John D. Rockerfeller.

Development is the process in which individuals increase their abilities and desires to satisfy their own needs and legitimate desires, and those of others. It is at least as much a matter of motivation, information, knowledge, understanding, and wisdom as it is of wealth. An individual's *level of development* is his current ability and desire to satisfy his own needs and legitimate desires, and those of others.

This definition requires clarification of the distinction between *needs* and *desires*, and of the meaning of *legitimate*. By a *need* we mean something that is necessary for survival, for example, food and oxygen. What is needed may or may not be desired; for example, a person who is unaware of his need for roughage does not desire it. On the other hand, as is obvious, a person may desire something he does not need.

By a *legitimate desire* we mean one the pursuit or fulfillment of which does not reduce the likelihood of other individuals fulfilling their needs or (legitimate) desires. This does not preclude interfering with a person who interferes with others' pursuit of their needs or legitimate desires.

Development, as we define it, is reflected at least as much in quality of life as in standard of living. To be developed is to have the desire and capacity to use effectively whatever one has to improve one's own quality of life and standard of living, and those of others.

It has become more and more apparent that continued economic growth of a nation may increase the standard of living, but not necessarily improve the quality of life. Many argue that at least some of the economically most advanced nations today are increasing their standards of living at the expense of quality of life. Many American Indians, for example, assert that although their assimilation into the surrounding white culture has brought them a higher standard of living, their quality of life has deteriorated.

All this is *not* to say that wealth is irrelevant to development or quality of life; it is very relevant. How much people can actually improve their quality of life and that of others depends not only on their motivation, information, knowledge, understanding, and wisdom, but also on what instruments and resources are available to them. For example, a man can build a better house with good tools and materials than he can without them. On the other hand, a developed man can build a better house with whatever tools and materials he has than a less developed man with the same resources. Furthermore, a developed man with limited resources can often improve his quality of life and that of others more than a less developed man with unlimited resources.

It should also be borne in mind that resources are more often *taken* than given. The more developed a person or an organization, the more resources he or it can find or develop. The more dependent one is on resources that are given, the less developed that person is. Put another way: resources are created by what man does with what nature provides. What nature provides is not a resource until man has transformed it or learns how to use it. The more developed man is, the more resources he can create or extract out of nature's offerings.

Because development involves an increase of ability (i.e., learning) and one person cannot learn for another, one person or organization cannot develop another. One can only encourage and facilitate that development. There is only one type of development: *self-development*. One who tries to induce development in another much act like a teacher, not a doctor. It is not so much a matter of diagnosis and prescription as it is of encouragement and facilitation.

Therefore, social systems cannot develop their members and other stakeholder, but *they can and should encourage and facilitate such development.* The development of such systems consists of *an increase in their desire and ability to encourage and facilitate the development of their stakeholders.* Effective planning for development requires a major reorientation of organizations or governments that assume they can develop their members. Development is not a matter of what an organization or government does, but of what it encourages and enables its members to do.

The quality of life (including work life) that a person can realize is the joint product of his development and the resources available to him. Although this implies that limited resources may limit improvement of quality of life, it does not imply that they limit development. Desires and abilities can increase without an increase of resources. Put another way: development is *potentiality* for satisfaction of needs and desires, not the satisfaction (quality of life or standard of living) actually obtained.

A limit is a point, line, surface, quantity, or other boundary that a variable cannot exceed. For example, nothing can move faster than the speed of light which, therefore, is a limit. The maximum speed at which an automobile can travel is also a limit. Nevertheless, we speak of speed limits which automobiles obviously exceed. Here "limit" denotes a quantity that one is not supposed to but can exceed. Such a limit is better referred to as a *constraint* or *restriction*.

Even the limit to the speed of an automobile need not limit its driver. He can take an airplane if he wants to travel faster or he might find something more desireable than travel. The effect of limits on purposeful individuals and systems can be evaded either by changing intent or by using better technology. A limited resource limits us only if we want to do something that requires more of that resource than is available and there is no suitable substitute in greater supply. A limited resources ceases to be limiting if our need for it decreases or *if we learn how to use it more effectively*; that is, *if we develop*.

Limits to and constraints on a social system's growth are found primarily in its environment; but the principal limits to and constraints on its development are found within the system itself. The key to unlimited development is freedom of choice that is limited only to those who do not limit the choice of others; that is, freedom that is exercised morally.

CONCLUSION

We have tried to show that there are three different ways of looking at and thinking about social systems: as mechanistic, organismic, or as social systems. We have argued that the first two types of modeling severely limit our ability to understand such systems, hence to control them. A mechanistic model yields a description of the actions and interactions of the parts of a social system. and, therefore, knowledge of its structure. It does not yield understanding of the behavior of either the parts of the whole. Organismic models of social systems lead to identification of the functions of the parts in the whole. The social model conceptualized a social system as a part of a larger purposeful system as well as a system with purposeful parts. It focuses on both the functions of the parts in the whole and of the whole in the larger containing system of which it is part. Therefore, it can yield understanding of both the behavior of the parts and the whole.

Mechanistic management strives for efficiency and tries to construct social systems that behave like machines, and to train people to behave like machine parts. This results in a bureaucratic structure that is capable of neither learning nor adapting. In an environment characterized by an increasing rate of change, uncertainty, and complexity, mechanistically managed organizations tend to become increasingly dysfunctional, to lose effectiveness, and often to die although burial may be postponed.

Organismic management strives for growth as necessary for survival. It conceptualizes the parts of a social system as organs with essential functions but no purposes of their own. Their only reason for existence is their service to the whole. Their environments are seen as purposeless and passive providers of necessary inputs to, and outputs of, organisms. Such organizations tend to be managed more permissively than bureaucracies, focusing on meeting assigned goals, leaving choice of means by which these goals are to be pursued to the parts that have responsibility for their attainment. They give more attention to efficiency than effectiveness. For them, planning consists of predicting a future environment believed to be beyond their control, and preparing for it.

Social-system management is concerned with development and tires to serve the purposes of the system, its parts, and its containing systems. There may be conflict between these levels or within them. Therefore, resolution or dissolution of conflict is one of management's principal responsibilities. A social system should be viewed as an instrument of those it affects. Its principal function is to encourage and facilitate their development. For management of social systems, planning should consist of designing a desirable future and inventing or finding ways of approximating it as closely as possible. Such management should attempt to maximize the freedom of choice of those it affects. Only from experience of choice can one learn, hence develop.

REFERENCES

Ackoff, R. L. and Emery, R. E., *On Purposeful Systems*. Seaside: Intersystems Publications, 1981.

Beer, Stafford, *Brain of the Firm*. Chichester: John Wiley & Sons, second edition, 1982.

Boulding, Kenneth, *Beyond Economics*. Ann Arbor: University of Michigan Press, 1970.

Buckley, Walter, *Sociology and Modern Systems Theory*. Englewood Cliffs: Prentice Hall, 1967.

March, J. E., and Simon, H.A., *Organizations*. New York: John Wiley & Sons, 1958.

Meadows, D. H., Meadows, D. L., Randers, J., and Behrens, W. W. III, *The Limits to Growth*. New York: Universe Books, 1972.

Naisbitt, John, *Megatrends*. New York: Warner Books, 1982.

Singer, E. A., Jr., *Experience and Reflection*. Philadelphia: University of Pennsylvania Press, 1959.

2
CULTURE AS STRUCTURE OF SOCIAL SYSTEMS

There are many things about the behavior of a social system that refer to the nature of the interaction between its elements rather than to the individuality of its members.[11] Each social system manifests certain characteristics and may retain these even if all its individual members are replaced. What characterizes a social system is not only its members, but the relationship of its members to one another and to the whole This is, of course, implicit in the definition of a system.[1] Some kind of linkage between the elements is presupposed if the aggregate is to be considered a system. The point of emphasis, therefore, is not the existence of a relationship, rather, it is to specify explicity the assumptions regarding the nature of these relationships.

These relationships, which in our conception denote the structure of a system, in turn depend on the nature of the bonds that link and hold the components of the system toegether. In this context there are fundamental differences between the nature of the bond in mechanical and in socio-cultural systems.

Whereas the elements of mechanical systems are *"energy-bonded,"* those of socio-cultural systems are *"information-bonded."* In energy-bonded systems, laws of classical physics govern the relationship existing between the elements. Therefore passive and predictable functioning of parts is a must, until a part breaks down. An automobile yields to its driver regardless of his expertise and dexterity. If a driver decides to run a car into a solid wall, the car will hit the wall without objection, but riding a horse presents a different perspective. It matters to the horse who the rider is, and a proper ride can only be achieved after a series of information exchanges between the horse and the rider. Horse and rider form an information-bonded system in which guideance and control are achieved by a second degree agreement (agreement based on a common perception) preceded by a psychological contract.

Buckley[7] provides a way of understanding this structural characteristic of socio-cultural systems by focusing on the organization and its dynamics based on the effect of information, as opposed to energy transmission. The socio-cultural system is viewed as a set of elements linked almost entirely by way of intercommunication of information. It is an organization of meanings emerging from a network of interaction among individuals.

The typical characterization of social structure as "the complex array of roles and statuses" has been challenged not only by Marxists with various understandings of Marx, but by a host of diverse writers collectively designated as "structuralist," of whom Levi-Strauss is generally taken to be a leading figure and influence.[12]

Ironically, this group of structuralists, influenced by a strong tendency to find a one-to-one relationship between structure and function, has in general treated structure as a dependent variable, and has defined social structures in terms of a single function. But in this context each one has used a different function. For example, Orthodox Marxists consider the production of goods as the prime function of society and thus define social structure solely in term of "mode of production." Lenin, despite accepting the primacy of production function, considers power as the key factor in determining social structure. Whereas Bogdanov[4] finds the generation and dissemination of knowledge the critical function and concentrates on "mode of organization." Levi Strauss on the other hand, sees the language and "myth" in the core of an invisible social structure. This fragmented view of social structures at best represents partial reality and is an over-simplification. This leads to the notion that the answer may not be in one or the other, but in a creative synthesis.

Further confusion with respect to social structure derives from a long standing traditional dispute over the dichotomy of objectivity and subjectivity in social theory. According to Peter Manicas:[12]

> "Objectivism and positivism in social theory see that social structures are real (objective), because they have real effect but end up denying if subtly that they are the product of active agents. Subjectivism and voluntarism in social theory, by contrast, maintain that social structures are constituted by active agents but deny if subtly, that social structures have determinative (objective) reality."

But the assertion that social structures are real is not inconsistent with the idea that they are produced by actions of social agents. Giddon's[10] notion of "structuration," which refers to the transformation of social structures as they are reproduced, recognizes the truth of both voluntaristic and deterministic conceptions of social structure. The fact that social structures pre-exist for individuals[5] and thus have a real effect and even coersive power over them does not mean structures cannot be transformed and reproduced by the purposeful actions of their members. This conception is compatible with the systemic view of coproduction, in which three factors — 1) initial conditions, 2) chance, and 3) choice (purposefulness) — together coproduce the process of change.

CULTURE AS STRUCTURE OF SOCIAL SYSTEMS

Furthermore, systems view denies one-to-one relationships between structure, function, and process; it considers each one as an independent variable. This means that a given function can be produced by different structures and a given structure might produce several functions, and vice versa.

However, these three independent variables are so interwoven that none can be understood in isolation. Together they define the whole or make the understanding of the whole possible.

The state of a social system, at any given time, is defined by the four independent variables of structure, function, process and the environment.

$$[ST = f_1(S, F, P, E)]_t$$

ST = State of the system
S = Structure: components and their relations (MEANS)
F = Function: output produced (ENDS)
P = Process: rules of transformation
E = Environment: uncontrolled variables

Accordingly, process of change of purposeful systems — systems having the choice of both ends and means — can be represented by

$$[P = f_2(ST, E, S, F)]_t$$

In this formulation we can represent the environment with the factor of *chance*, the structure (choice of means) and function (choice of ends) by the factor of *choice* and the state of the system by its *initial conditions*.

Note that probabilistic transformation is the special case of fixed structure and fixed function (no choice situation), in which the environment becomes the only relevant factor. On the other hand, deterministic transformation represents the case of fixed structure, fixed function, with no environmental effects.

Finally, in systems view social structure is seen as a temporary expression of relations between a system's components and its environment. These relations can be observed only in the context of its total functions — generation and dissemination of knowledge, power, wealth, values and beauty — and the process of its change. Therefore, instead of trying to describe social structure in terms of being, we have to understand it as a process of becoming.

This concept of social structure is best manifested in the culture of social systems. To elaborate on this and to clarify the meaning of information bondedness, we need to examine the concepts of culture and of social learning. That will be the aim in the rest of this chapter.

CULTURE

Image building and abstraction are among the most significant characteristics of human beings, allowing them not only to form and interpret real images of real things, but to use these to create images of things which may not exist. Man feels hunger, observes the fleeing prey, and realizes his inability to capture it. After discovering other related objective realities (wood, stones, etc.), he thinks about and eventually creates a subjective image of a tool, one yet to be, that would help him secure his food. Transformation of this subjective image into an objective reality results in the *bow and arrow*, which in turn will be a co-producer of yet another image, and so on. This dialectic interaction between objective and subjective realities lies at the core of a synthesizing process responsible for the dynamic development of human societies.

As a prerequisite to survival it has always been necessary for man to observe and understand events that are constantly occurring in his environment. He does this so that he may utilize favorable opportunities and be prepared for antagonistic events. But understanding scattered phenomena in isolation, although necessary, is not sufficient for man to relate to his environment. Therefore, an additional struggle to find a logical relationship among these isolated findings impels him to synthesize this fragmented information into a unified, meaningful mental image and eventually into a world view.

Co-produced by the environment and man's unique process of creativity, the image establishes a link between man and his environment. It consists of a system of assumptions (possibly unconscious) regarding the nature of spatio-temporal-causal realities in addition to a concept of values, aesthetics, and finally his perceived role in the environment. A considerable part of this image or mental model of the universe is shared with others who live in the same social setting. The rest remains private and personal.[6] It is the shared image that constitutes the principal bond among the members of a human community and provides the necessary conditions for any meaningful communication. The extent to which the image of an individual coincides with the "shared image" of a community determines the degree of his membership in that community. It is the "shared image" that we refer to as the *culture* of a people. It incorporates their experiences, beliefs, attitudes, and ideals and is the ultimate product and reflection of their history and the manifestation of their identity — man creates his culture and his culture creates him.

It is here that the key obstacles and opportunities for development are found, the collective ability and desire of the people to create the future they want. There-

CULTURE AS STRUCTURE OF SOCIAL SYSTEMS 21

fore, human culture with all its complexity, ambiguity, and manifold potentialities stands at the center of a process of change. This process cannot be understood except against the background of the culture of which it is a part, which it builds upon and reacts against. So much is this so that the success of individual actions invariably depends on the degree to which they penetrate and modify the "shared image."

SOCIAL LEARNING

The potentiality and vitality of the culture, and thus that of the social structure, lies in its creative ability to meet the challenges of continuously emerging desires and ideals. This process demands conscious and active adaptation, not a passive acceptance of events. It is a struggle for the creation of new dimensions, appreciation of new realities and finally, enrichment of the common image. It is a learning process that entails coordinated changes in motivation, knowledge, and understanding throughout the social system. Although social systems learn through their members who adjust their world views or mapping of reality by observing the actual or potential results of their actions, social learning is not the sum of the learning of each member. Socio-cultural systems manifest greater inertia and resistance to change than their individual members.

Just like a high-level computer language that provides default parameters when programmers fail to provide one, the culture of a social system provides default values when actors fail to choose one explicitly. For example, if a man does not decide explicitly what kind of father he wants to be, the culture decides for him. The problem with this is that the implicitness of the underlying assumptions prevents actors from questioning their validity; therefore, they usually remain unchallenged and become obsolete. Furthermore, actors, by repeated use of default values, tend to forget that they have a choice and treat such values as "realities out there," undermining the fact that those "realities" will remain out there so long as no one is willing to challenge them.

Fear of rejection and a strong tendency toward conformity among members of a social system is another important obstruction to social change. An example is the experience of dry county whose constituents were about to vote on the issue of the alcohol ban. A prevote survey indicated that 75% of the voters favored abolishing the ban. However, individual voters thought the majority wanted to keep the county dry. When the votes were tabulated, 60% of the voters had voted to keep the county dry. Not surprisingly, after the result of the survey were published prior to next vote on the issue, the county went wet with a 65% majority.

Finally, the inertia of a culture, is manifested in the fact that public and private images act as filters, developing a selective mode of reception. This tunes the receptors for particular messages. Those consistent with the image are absorbed and further reinforced, while contradictory and antagonistic ones have no significant effect. This phenomenon, although an impediment to change, acts as a defense mechanism and structure-maintaining function. Furthermore, since truth is commonly identified with simplicity and comprehensibility, what one does not understand is simply rejected as false. A high level of specialization in science moves it further away from the common image, creating a small, isolated subculture. This reduces the needed influence of science on the behavior of the public at large. This is the dilemma of the democratic process and remains its main challenge.

Ironically, it is frustration that usually acts as the prime producer of change, the frustration of encountering a series of events for which prevailing mental models can no longer provide convincing explanations, the frustration of facing an increasing number of dilemmas which cannot be dissolved with existing sets of assumptions.

It is at this point that opportunity arises for different ideas to challenge the common image. Struggle between conflicting views will force the confused system to oscillate, causing further frustration. This oscillation will continue until the system appreciates the new dimensions involved and develops the capability to synthesize at a higher level. This finally leads to the restructuring of the common image and elaboration of the social structure, making it more compatible with the values and emerging realities of the new era. However, appreciation of new dimensions and orderly transformation of social systems require an ability for second-order-learning, which must be distinguished from first-order-learning.

Consider a choice model in which actors are to choose among several courses of action. This choice model is formed by what actors collectively believe are the possible courses of action available to them. Inclusion or exclusion of alternatives in the choice model is not arbitrary. The choices in the set usually share one or more properties based on an explicit or implicit set of assumptions or constraints produced by the actors' previous experience with similar situations. In this context, first-order-learning represents a quantitative change. It is a revision of probabilities of choice, modifying parameters, in a fixed structure. Underlying assumptions governing the selection of alternatives remain unchallenged.

Second-order-learning, on the other hand, involves the challenging of assumptions. It represents a qualitative change that results in a re-identification of the available set of alternatives and objectives. This redefines the rules for first-order-learning.[3] Unfortunately, in societies polarized by antagonistic and rigid ideologies social transformation takes place by a violent change of phase (a cusp). Retrieval from such a situation is often extremely problematic, since the relationship between members is irreparably damaged as happens in societies which are thrown into a perpetual state of civil disorder.

In this context, ideologies, in any form or type, represent a profound obstruction for second-order-learning. Of course an ideology should not be mistaken for a cause or a vision. The significant and common characteristic of all ideologies is a claim for ultimate truth with a pre-defined set of ends and means. Underlying assumptions are not to be questioned by true believers. This makes them incompatible with the requirements for second-order-learning.

The critical issues of qualitative change and the need to deal more effectively with social pathologies demand incorporation of second-order-learning in social systems. This requires the creation of a new mode of organization in the form of an ideal-seeking-system, in contrast to an ideal-state. This warrants further clarification.

Throughout history there have been repeated attempts to fashion human societies in accordance with some sort of idealized image. This has been done by prophets, philosophers, social reformers, and in recent times by the state apparatus in more than one country.

In all cases, these ideals have been defined by human authorities who have attempted to legitimize their authority by means of an ultimate authority, such as Science or God. But the identification of the ideal state with an ultimate authority precludes freedom to change. This is the essence of social pathology as we defined it.[9]

Within this framework of acceptance of an ideal state defined by an ultimate authority, it is possible to distinguish between two approaches:

The first approach consists of specifying a detailed and comprehensive set of rules of conduct for individual behavior which, if followed by all members of society, would automatically lead to the emergence of the ideal state.

Prime examples of this approach can be found in religious attempts to reform as well as the utopian socialism of the 19th Century. In the name of God or ultimate truth, the objective of this approach has been the creation of a "new man" who will better conform with the mechanistic image of the ideal society. Ironically the repeated failures in changing the "nature of man" into a preprogrammed robot has not reduced the commitment of "true believers" in their pursuit. On the contrary, enjoying a phenomenal capacity for denial and self-deception, they blame the weakness of man for the failure, and see an urgent need for total control by establishment of a totalitarian order. No wonder that in the 20th Century, utopianism is identified with authoritarianism in the minds of many.

The second approach is characterized by the struggle to create a new social structure based on the assumption that man is solely the product of his environment, and that his behavior is basically a reaction to it. This school of thought maintains that a certain political/economic order, based on the "objective realities of social existence," is necessary and sufficient to bring about the desired mode of conduct, leading to the realization of the ideal state.

Scientific socialism, which represents the first attempt of this approach, degenerated in practice into the first type. This happened once it was realized that the set of necessary and sufficient objective conditions to bring about the desired pattern of human conduct was not predicted in sufficient detail by the founder-theoreticians of the movement or their successors. Following the disillusion with finding the proper structure and the failure of the established order (Weberian bureaucracy) to produce the expected outcome, the leaders of those countries that attempted so-called "scientific socialism" found themselves with only the alternative of defining rules of conduct and resorting to the old idea of creating a "new man." This is the same concept they had once rebelled against.

The fundamental problem with both of these approaches, which despite their apparent differences result in the same practical consequences, is in their misconception of the nature of the ideal state and the processes which bring it about. They both contend that:

1. There is one and only one end (ideal state) predefined by an ultimate authority (God or Science).
2. The ideal is not only attainable, but the movement toward it is also inevitable.

The scientific justification for this position is provided by the historical determinism of the mechanistic worldview, or the equifinality of the organismic worldview. In the latter case, although the end is predefined and fixed the social system has the choice of means. Therefore, by selection of efficient means it can expedite the attainment of the promised end.

The inevitability of the final state, and its independence from the generating processes, leads to the notion that "the end justifies the means." It is assumed that the seizure of power by the chosen class or group who is supposed to be the sole beneficiary of the ideal state, is the precondition for its realization. Therefore the struggle for political power becomes the main preoccupation of true believers in a predetermined ideal state. Furthermore, a profound confusion about the nature of conflict and contradictions in social context, the dichotomic treatment of dialectical realities, results in a win/lose struggle, and justification for elimination of opposing groups or classes.[8]

But, ideals, in systems view, are regarded as dynamic and changing over time. The image of an ideal state, defined by the members of the social system, reflects the spatio-temporal realities (here and now) of the particular historical moment, and thus is alterable even before being approached (moving target).

The systems view, by rejecting the concept of a single absolute ideal, accepts the principle of multifinality as opposed to equifinality. Equifinality states that the

CULTURE AS STRUCTURE OF SOCIAL SYSTEMS

final state is fixed and predetermined although attainable by different paths and from different initial conditions. Death of animate subjects and the Communist state in the Marxist view are examples of such final states. Multifinality on the other hand, holds that there may be several final states, each attainable by different paths. Even starting from the same initial conditions social systems do not necessarily attain the same final state.

While not denying the influence of the past in shaping the future, systems thinking does not regard the past as the sole determinant, but only a co-producer of the future. As Ackoff[2] puts it: "the future is not completely contained in the past, much of it is yet to be written."

By considering man as a purposeful system, with the choice of both ends and means, the systems view rejects efforts aimed at degrading him to the level of a robot. Behavior of a purposeful individual is regarded as a *response* rather than a *reaction* to his environment. (An event in the environment is both necessary and sufficient for a reaction. It is only necessary but not sufficient for a response.) The recognition of the element of choice in the behavior of social systems leads to the belief that these systems have the capability to select their own future and successively *approximate* it by choosing the appropriate means.

In the systemic view every phenomenon is the end result of one or several processes. Thus, in order to bring about the desired end it is necessary to choose appropriate and consistent processes for its attainment. In particular, means which negate the end, or are in conflict with it cannot be effective in bringing it about. Creation of a hero to champion the cause against heroism is a self defeating proposition. There is no "just" dictator, whether it be a religious "saint" or an intellectual appointing himself as the guardian of the working class. Unlike the concept of equifinality in the organismic model, where the final state is attained irrespective of the chosen path, for a social system the end state is not independent of the generating processes. The means are themselves among co-producers of the end, directly influencing the essential qualities of the resulting phenomenon.

To summarize.

The structure of a system defines its components and their relationships. These relationships, in turn, depend on the nature of the bonds that link and hold the components together. However, in contrast to physical systems which are energy bonded, social systems are information/knowledge bonded. As the energy level determines the mode of organization in physical systems the knowledge level defines it for social systems. Thus the role of knowledge in social systems is analogous to that of energy in physical systems. But unlike energy, knowledge is not subject to the "law of conservation. The ability to learn and create knowledge enables social systems to constantly re-create their structures. But the knowledge level of a social system is not the sum of the knowledge of its individual members. It is the shared knowledge, or shared image, as manifested in the culture that defines the knowledge level of a social system. Therefore, proper transformations of social structures require a cultural change, creation as well as dissemination of knowledge. This in turn demands a capability for second-order-learning.

Finally, it is essential to note that by emphasizing the learning process, the systems approach to idealization is not to design an *Ideal State* but an *Ideal-Seeking-System*.

Design of an ideal-seeking-system is in fact at the core of any effective means for dealing with social pathologies, and realizing the full potential of a purposeful organization; that is,

- participative process which enables the members of a social system to collectively define and redefine their desired futures and relate their roles to the totality of the system of which they are a part;
- a learning and adaptive system that is *able* and *willing* to alter its course at any time in recognition of emerging values and new realities;
- a pluralistic social setting which will encourage and facilitate questioning of sacred assumptions and challenging the implicit default values of the culture.

REFERENCES

1. Ackoff, R. L., "Towards a System of Systems Concepts," *Management Science*, Vol. 17, No. 11 (1971).

2. Ackoff, R. L., *Redesigning the Future*, John Wiley (1974).

3. Ackoff, R. L. and Vergara, E., "Creativity in Problem Solving and Planning: A Review," *European Journal of Operational Research*, 7 (1981).

4. Bogdanov, A., *Essays in Tektology*, Translated by George Gorelik, Intersystems Publications (1980).

5. Bhaskar, R., "The Possibility of Naturalism," *New Jersey Humanities Press* (1978).

6. Boulding, K. E., *The Image*, Ann Arbor (1956).

7. Buckley, Walter, *Society as a Complex Adaptive System in Systems Research for the Behavioral Scientists*, edited by Buckley, Aldine Publishing Co. (1968).

8. Gharajedaghi, J., "Social Dynamics, Dichotomy or Dialectic," *Human Systems Management*, 4 (1983).

9. Gharajedaghi, J., "Obstructions to Development," *Human Systems Management*, 4 (1984).

10. Giddins, A., "New Rules of Sociological Method," *Hutchinson of London* (1976).

11. Laszlo, E., *Systems View of the World*, George Braziller (1972).

12. Manicas, P., "The Concept of Social Structure," *Journal for the Theory of Social Behavior*, 10, 2 (1980).

3
SOCIAL DYNAMICS:
DICHOTOMY OR DIALECTIC*

Social change is the result of an interacting network of various social tendencies, and relationships of many different kinds. There seems to be no single "leading factor" which is always in the forefront; and the relationship cannot always be reduced to a simple dichotomy — of either conflict or cooperation. In a given system interactions between opposing tendencies can take many forms including: conflict, competition, cooperation and collaboration, which may coexist simultaneously. Actors in a social system may cooperate with regard to one pair of tendencies, compete over others and be in conflict with respect to different sets at the same time.

However, the "HOW" and "WHY" questions of social transformation, the concept of conflict, its role in the process of change, and, finally, the challenge of converting a win/lose struggle to a win/win strategy, are still among the major concerns for designers of social systems.

In this context, it is my contention that the continued treatment of multidimensional realities as unidimensional concepts — simple dichotomies or continua — is not only at the core of the confusion and dilemmas, but also represents a fallacy, into which many people of good will often fall. The point is that sets of opposing tendencies — assumed to be the prime producers of change — are in fact, two sides of the same coin; they coexist and interact continuously.

In this work I will make an attempt to organize my understanding of social dynamics by elaborating on the concepts of (1) AND/OR relationships, (2) change of phase and mode of organization, and (3) integration and differentiation. Then, I will speculate on a framework which I hope will be useful and instrumental in "dissolving" the dilemmas generated by win/lose struggles which presently dominate the social environment.

---

*This paper originally appeared in *Human Systems Management*, Vol. 4, 1983, Elsevier Publisher B.V. (North-Holland)

1. AND/OR Relationships:

In viewing AND/OR relations in the context of social dynamics we are concerned with processes not states. We are dealing with opposing tendencies not opposing states. This distinction is central for understanding the concept of dialectic and the processes of social change. Although a person cannot be alive *and* dead at the same time, she/he can display tendencies toward both.

A purposeful individual has different ways of dealing with conflicts, she/he can either solve, resolve, absolve or dissolve them.[1]

To *solve* a conflict is to select a course of action that is believed to yield the best possible outcome for one side at the cost of the other, a win/lose struggle.

To *resolve* a conflict is to select a course of action that yields an outcome that is good enough and minimally satisfies both of the opposing tendencies, a compromise.

To *absolve* a conflict is to wait it out, to ignore it and hope that it will go away.

Finally, to *dissolve* a conflict is to change the nature and/or the environment of the entity in which it is imbedded so as to remove the conflict.

Selection of any/one of these courses of action depends on how the relationships between opposing tendencies are formulated. There seems to be at least three ways in which these relationships are conceived.

First, there is the conceptualization of conflicting tendencies as two mutually exclusive, discrete entities. Here conflicts are treated as dichotomies which are usually expressed as "X" or "NX". This represents an "OR" relationship,[2] thus a win/lose struggle which calls for a *solution* of the conflict. The loser is usually declared wrong and eliminated.

Figure 1. Dichotomic treatment of conflict.

Second, there are unidimensional conceptions of conflict situations. Here conflicting positions are formulated in such a way that they can be represented by a continuum. Here there seem to be many shades of gray between black and white. This calls for a compromise or *resolution* of the conflict.

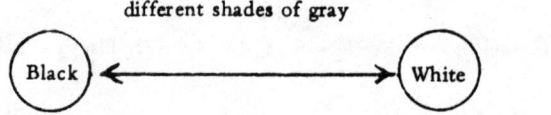

Figure 2. Unidimensional conception.

SOCIAL DYNAMICS: DICHOTOMY OR DIALECTIC   31

A compromise is thus a mixture of ideas of both poles of tension. It might seem to be a kind of integration. However, this is usually quite superficial. It usually contains elements of the two extremes but it does not provide a new framework that can encompass both poles.

Third, there are multidimensional conceptions of conflict, in which interaction between opposing tendencies is characterized by an "AND" relationship.

In contrast to *OR*, the *AND* relationship recognizes the mutual interdependence of opposing tendencies. In this conception, the opposing tendencies not only coexist and interact, but also form a complementary relationship. A complement is that which fills out or completes a whole.[3] The complementaries are not necessarily a pair. Many complementaries may coexist relfecting the essence of co-production and the producer-product concept.[4] This formulation requires a *dissolution* of the conflict.

Transforming a conflict seen as unidimensional into one that is seen as multidimensional is a dialectical process. In this interpretation of dialectic, which corresponds to the systemic concept of co-production, each one of the opposing tendencies is represented by a separate dimension, resulting in a multidimensional scheme. (Figure 3.2)

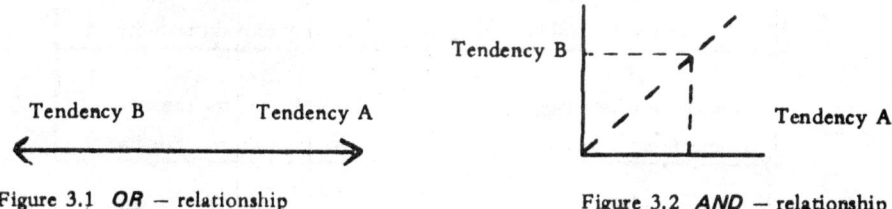

Figure 3.1 *OR* — relationship   Figure 3.2 *AND* — relationship

In the unidimensional conception of conflict — the conflict situation is formulated in such a way that a gain for one tendency is invariably associated with a loss for the opposing side, a zero-sum-game. (Figure 3.1) But the multidimensional conception of conflict characterizes a non-zero-sum situation, in which a loss for one side is not necessarily a gain for the other. On the contrary, this formulation permits both opposing tendencies not only to coexist but also to increase or decrease simultaneously. Therefore, *lose/lose* as well as *win/win* in addition to *win/lose* struggles are strong possibilities.

Using a multidimensional representation one can see how the tendencies previously considered as dichotomies can interact and be integrated into something quite new. The addition of new dimensions makes it possible to discover new frames of reference in which opposing sets of tendencies can be interpreted in a new ensemble with a new logic of its own.

Note that in classical logic, contradictions are also considered to be relative to a domain; adding a new dimension expands the domain and converts the contradictory pairs to contrary ones. Therefore

> "The usual dichotomy of "x" or "not x" never seems to display the general, because neither of the above is always so prominent an aspect of the social system"[5]

By going a step further we have the possibility of converting contradictions to alternatives.

For example: Consider the concern for stability (morphostasis) and change (morphogenesis). A unidimensional conception of these tendencies and subsequent classification of social theories based on their orientation toward one *or* the other, as done by Burrell and Morgan,[6] will lead to a dichotomy of "sociology of regulation"[7] versus "sociology of radical change."[8] But, on the other hand, if we recognize that the interaction between the tendencies for stability and change are among important dialectical phenomena, then their representation by a two dimensional scheme will lead to four categories instead of two.

		LOW STABILITY	HIGH STABILITY
Change	High	(2) High concern for change / Low concern for stability	(3) High concern for change / High concern for stability
	Low	(1) Low concern for change / Low concern for stability	(4) Low concern for change / High concern for stability

Note that two previous dichotomous categories are represented only in quadrants *2* and *4* as special cases in each of which a high concern for one tendency is coupled with low concern for the other and vice-versa.

In the individual context, the work of Gerald Gordon[9] and his colleagues regarding the study of the factors conducive to innovation, provides additional supporting evidence for an "AND" relationship. Gordon and his colleagues see the following two abilities as complementary to an individual's propensity to innovate: the ability to differentiate between objects which seem to be similar — and the ability to find similarities between seemingly unrelated matters.

SOCIAL DYNAMICS: DICHOTOMY OR DIALECTIC

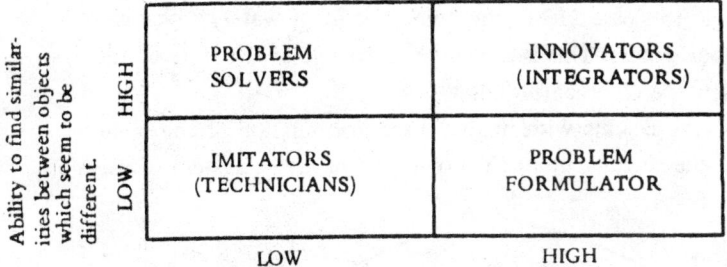

On the other hand, the constant struggle between groups of people who somehow see quite different "clear and urgent" necessities in dealing with different aspects of social realities — the urgency of production versus that of distribution, the desire to protect the rights of victims versus the rights of the accused, individual choice versus collective choice[10] — is the manifestation of the need to develop a new framework, which will allow the creation of a win/win environment and dissolution of the conflict situation.

Churchman's concern with the "environmental fallacy," Boulding's rejection of suboptimization ("the name of the devil is suboptimization") and Ackoff's concept of "separately infeasible parts making a feasible whole" are reflections of the same principle.

But, dissolving a conflict, as mentioned before, requires a change in the nature or the environment of the entity in which conflict is imbedded. This is a qualitative change reflecting a new mode of organization for the entity or its environment.

The next section tries to clarify this point.

2. The Change of Phase and the Mode of Organization

One of the commonly accepted principles of dialectic is that "A Quantitative change will result in a Qualitative change." This statement does not mean that an increase in quantity of a given variable will result in a qualitative change in the variable itself. Qualitatively there is no difference between one pound or a billion pounds of sugar. However, when the state of a system depends on two or more variables a quantitative change in one variable alone, *beyond a critical point*, will result in a change of phase in the state of the system. This change is a qualitative one, representing a whole new set of relationships among the variables coproducing the state of the system.

Suppose the state of a system (water) depends on two variables (pressure and temperature). Under constant pressure (atmospheric) if the temperature of water increases (a quantitative change) from 33°F to 211°F, no qualitative change in the state of the system (liquid water) is observed. However, when an increase in the temperature exceeds the critical point (in this case 212°F, the boiling point of water) a change of phase from liquid to vapor occurs. This is a qualitative change in the state of the system, representing a new mode of organization.

Catastrophe Theory,[11] dealing with mathematical formulation of the same principle also reveals that at the singular point ($Y''=0$) the state of the system displays a catastrophic behavior (a cusp).

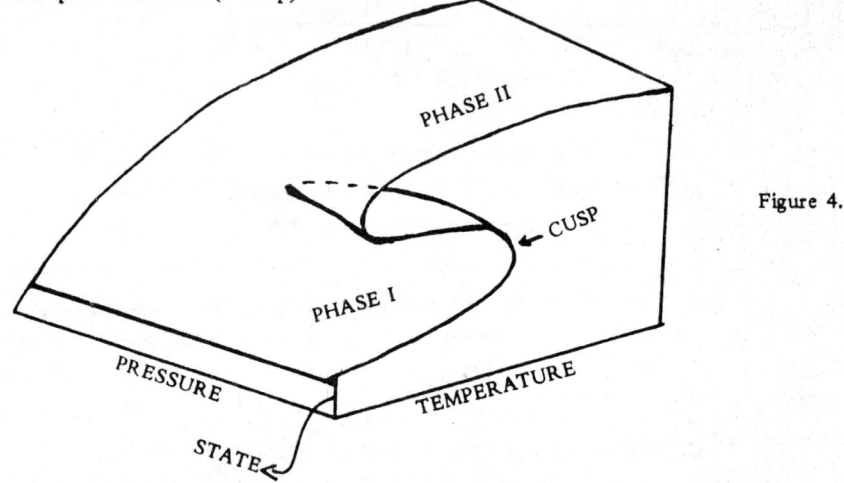

Figure 4.

The formation of the cusp signifying the change in phase (Figure 4) is, of course, not inevitable. It depends upon the behavior of the other variables affecting the system state. For example, in the case of water discussed above, the critical point at which vapor is formed changes with variation in the pressure. If both temperature and pressure are raised simultaneously, the cusp is delayed, and catastrophic change of phase does not take place at 212°F. Extension of this physical analogy to social phenomena requires a note of caution, and appreciation of the following distinction is especially critical.

As I have argued elsewhere,[12] in contrast to physical systems which are enery bonded, social systems are information/knowledge bonded. As the energy level determintes the mode of organization in physical systems, the knowledge level defines it for social systems. Therefore, the role of knowledge in social systems can be said to be analogous to that of energy in physical systems.

But the significant point, however, is that knowledge, unlike energy, is not subject to the first law of thermodynamics (the law of conservation of energy). One does not lose knowledge by sharing it with others. On the contrary, its dissemination increases the knowledge level of the social system. This important capability — creation of knowledge — will be shown to enable a social system to constantly recreate its

SOCIAL DYNAMICS: DICHOTOMY OR DIALECTIC

structure, and makes it possible that a change of phase in social systems could take place without a significant sign or a catastropic cusp.

Nevertheless, it can be said that the Change of Phase in physical systems (solid-liquid-gas) is analogous to the change in mode of organization in the social systems (feudalism, capitalism, and socialism), but the analogy ends there.

In the social context, a quantitative change in a variable also produces a qualitative change in the state of the dependent system. However, this change is produced through a reformulation or, more precisely, a reconceptualization of the variables involved. To explain this further let us look at a related concept, that of typology. Social typologies can be used to clarify why different modes of organization display quite distinct behavior regarding the same social phenomenon, providing one is aware of their underlying assumptions and limitations.

A proper way of developing typologies, which corresponds with my intentions here, requires that the relevant variables, which together define the state of the phenomenon under study, be identified and each one conceptualized as a separate dimension.

A dimension represented by an arrow, usually, is used to reflect a quantification of a variable on a given scale – it measures a characteristic (or behavior) specified by the operational definition of the variable involved. Segmentation of this scale into two regions of *low* and *high* is usually based on an assumption that the low or high value assigned to the variable will have a significant impact on the behavior of the system which is coproduced by the variable.

In this context the point of distinction between low and high is not arbitrary.

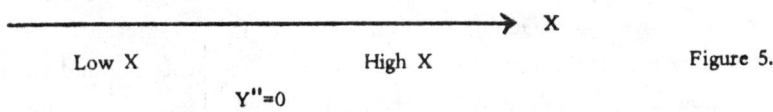

Low X   High X   Figure 5.
$Y''=0$

It signifies the level at which the behavior of the dependent system is qualitatively affected – corresponding to the critical point in phsycial phenomena (change of phase). This is to say that if variable income has an effect on the behavior of an individual, there seems to be a critical level of income at which a change in lifestyle is observed, and the behavior of the individual is qualitatively affected.

Although locating the singular point ($Y''=0$) in the social context is the major problematique of this conception, even in the absence of sufficient knowledge, the awareness of its existence and its implications is useful in understanding the nature of the qualitative differences observed in the behavior of the system based on the degree of the emphasis (or concern) put on one or the other dimension.

Now, recall the previous example, concern for stability and change. Note that the behavior of the system not only depends on two variables but also on the type of the relationship (AND/OR) which exist between them.

For example, the interaction of a high-concern-for-change with high-concern-for-stability produces a completely *different* mode of behavior than the one produced either by a high concern for change coupled with low concern for stability or the one produced by a high concern for stability coupled with low concern for change.

	LOW	HIGH
High Change	(2) LOW/HIGH Radical	(3) HIGH/HIGH (Ideal-seeking-system)
Low	(1) LOW/LOW Anarchy	(4) HIGH/LOW Conservative

Stability

The High/High (Quadrant 3) represents the behavior of an Ideal-seeking-system, searching for stability through change. While the Low/High (Quadrant 2) reflects a radical system interested in a change at any price; it can be reactionary or progressive depending on the direction of the change sought. Sometimes the action of a radical system is motivated by excitement of destruction without any concern for its consequences. The High/Low (Quadrant 4) on the other hand, represents a conservative state, preferring the status-quo, therefore, a concern for regulation and compromise. But the Low/Low (Quadrant 1) is anarchy with low concern for change coupled with low concern for stability, opposed to any form of government, interested only in "autopoiesis," which is a form of self-regulation.

Therefore, with different combinations of levels of concern (low or high) of the various tendencies, different modes of organization will emerge. Each mode represents a new system whose behavior can only be understood in its totality.

The typology of management style developed by Blake and Mouton[13] as shown in the following table also underlines this conception, by demonstrating that although "1,9" and "9,9" styles both reflect a high concern (9) for people, the manifestations of these concerns are different.

SOCIAL DYNAMICS: DICHOTOMY OR DIALECTIC

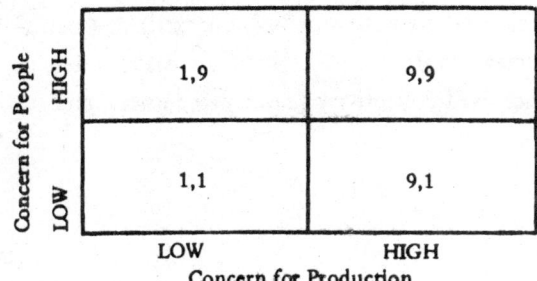

The "1,9" is a paternalistic leader, whose concern for people is basically a concern for their weakness. Therefore he assumes a protective role. While the "9,9" is a leader whose main concern for people stems from a respect for their ability. This requires a different role, the role of motivator and facilitator.

Now, to sum up this understanding of the social transformation and to generalize the concepts presented so far, we turn to a brief discussion of differentiation and integration.

3. Differentiation and Integration

The existing patterns of social transofmration and the subsequent creation of successive modes of organization, involve (1) an active process of the generation and dissemination of knowledge, (2) a process of learning and adaptation, (3) the creative process of discovery of a new dimension with all of its implications, and finally (4) the painful process of reconceptualization, reformulation and integration of all the variables involved in a new ensemble with whole new relationships and characterisitcs of its own.

Development of the social systems is unlike the physical pattern, which is conceived to be a unidimensional movement toward increased complexity in the structure of the matter. And it is unlike the biological evolution, which reflects a two-dimensional movement toward complexity and order. It is conceived to be at least a three-dimensional phenomenon of *purposeful* transformation in the direction of increased *integration* and *differentiation*.

The appreication of the interdependence of the following opposing tendencies — which have been conveniently grouped under the two generic terms of differentiation and integration — is the key to understanding the critical processes in the development of social systems.

Differentiation represents an artistic orientation with emphasis on intrinsic (stylistic) value systems, signifying tendencies toward such things as: increased complexity, increased variety, increased individual autonomy (individual choice), and morphogenesis (creation of new structure).

Integration represents a scientific orientation with emphasis on extrinsic (instrumental) value systems, signifying tendencies toward such things as: increased order, increased uniformity and conformity, increased collectivity (collective choice), and morphostasis (maintenance of structure).

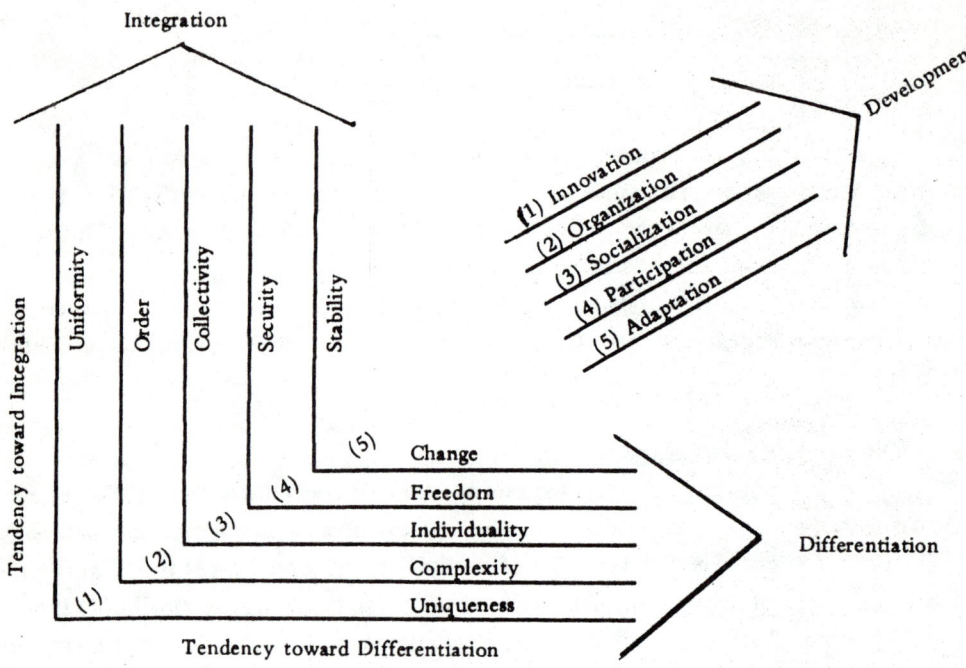

Figure 6.

In the above graph, the third dimension — purpose — has been omitted for simplicity. It will be discussed later in terms of agreement or disagreement between social actors, on the compatibility of their ends, means or both.

The emerging tendencies shown in the graph, that is the tendencies toward innovation, learning and adaptation, socialization (parity), participation, and organization, cannot stand alone. Together they form the whole, and coproduce a process called development. In addition, these tendencies are involved in and constitute the developmental processes for all of the social functions. These functions are the generation and dissemination of knowledge, power, wealth, value and beauty.[14] Thus, a holistic view of societal development, requires that, all of the five social functions develop interdependently, utilizing all of the five complementary processes outlined above.

SOCIAL DYNAMICS: DICHOTOMY OR DIALECTIC

The concept of "overdetermination" introduced by Louis Althusser[15] is an indication that some of the "new left" recognizes the interdependency and multidimensionality of social functions. So much so that Althusserian interpretation of dialectic is in line with and a reflection of the systemic concept of coproduction. As Boulding points out, the systemic vision of social dynamics is

> ". . . unfriendly to any monistic view of human history that seeks to explain it by a single factor, whether this is a materialistic interpretation, as in the case of [classical] Marxism, a simple theistic interpretation, as in biblical judaism, or an eschatalogical interpretation in terms of some simple denouncement . . . The simple rhetoric of class struggle and revolution, therefore must be regarded as an essentially minor element in the ongoing process of human and societal evolution, although it is sometimes important as a special case under particular circumstances."

Depending on the characteristic of a given culture, the orientation of its organizing group, and the relative emphasis put on integrative or differentiative tendencies, varieties of different modes of organization are observed in a social system.

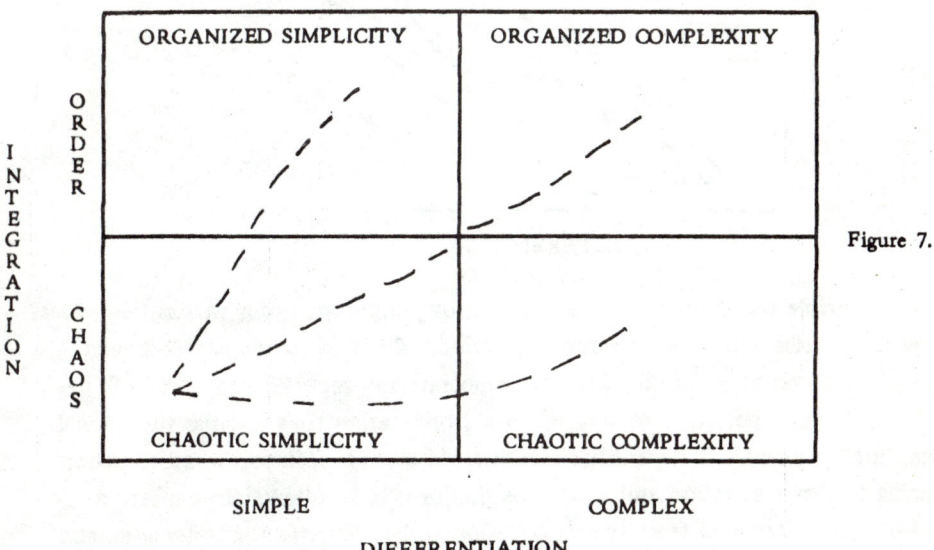

Figure 7.

A social system can move from a state of *chaotic simplicity* toward *organized simplicity* which in comparison with chaotic simplicity is a state, containing less variety, more uniformity, stronger bond among elements, produced by emphasis on integration at the cost of differentiation. It can also move toward *chaotic complexity*, a state with increased variety, reduced wholeness, increased diffusion produced by increased differentiation at the cost of integration. Or it can move toward *organized complexity*, signifying a higher level of organization achieved by movement toward complexity and order concurrently. Note that movement toward complexity and order is the essence of the negentropic processes in living systems. This mode represents the systems view of organization, that is, purposeful systems with purposeful parts and information/culture bonded systems with the capacity for structure creation as well as structure maintenance.[17]

Furthermore, one can speculate that for a given culture there is an upper and lower limit for integrative and differentiative processes respectively. Within the boundary of a given culture the variety of different orientations can be observed. Existence of "left" and "right" in every social group and political party is the manifestation of this phenomenon.

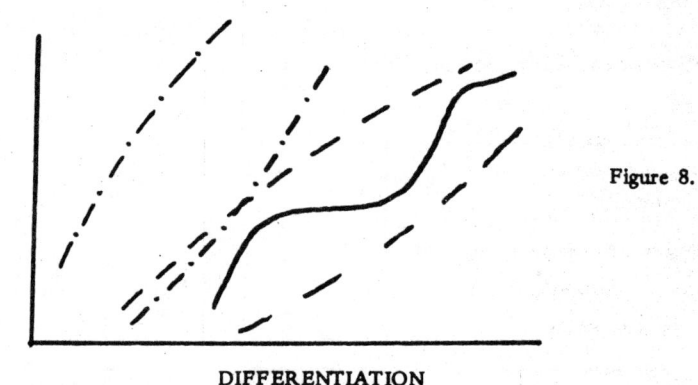

Figure 8.

In a flexible social system oscillations of low amplitude occur within the critical cultural boundaries without disruption. As periodic shifts of government between labor and conservative parties in U.K., or democrats and republicans in the U.S. demonstrate. But in a polarized society, if a rigid orientation tries to cross the critical cultural line, a violent and destructive reaction will move it back to the other extreme, producing further frustration and greater oscillations, and, finally, cusping into a change of phase. Retrieval from such a situation is then extremely problematic, and the relationship between members is irrepairably damaged as happens in societies which are thrown into a perpetual state of civil disorder.

Therefore corresponding to every level of differentiation there exists a minimum required level of integration below which a system would disintegrate into chaos. Conversely, higher levels of integration require higher degrees of differentiation in order to avoid oppression. This leads to the conclusion that the periodic oscillations and disruption witnessed in a social system are basically produced by dichotomous treat-

SOCIAL DYNAMICS: DICHOTOMY OR DIALECTIC

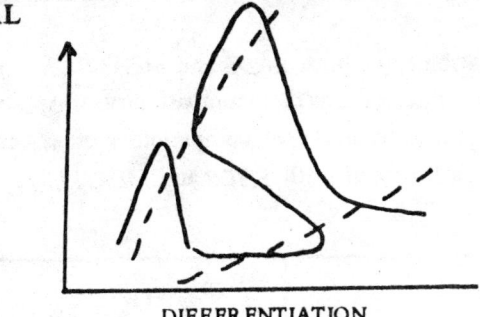

Figure 9.

ment of dialectical realities.

To remedy this situation, we need to understand how a unidimensional formulation of conflicting tendencies can be transformed into a multidimensional one so that a feasible whole can be coproduced by infeasible parts. The remaining parts of this chapter addresses this question.

4. Distinction Between a Dialectic and a Dichotomy

So far, we have argued, that the *AND* relationships denotes a dialectical *interaction* between opposing tendencies, while the *OR* refers to dichotomous *reaction* between them. Now the question is: under what condition are opposing tendencies dialectical and when do they form a dichotomy?

In understanding the distinction and implications of the dialectic and dichotomy in the social context, the concept of purposeful systems developed by R. Ackoff[18] is central.

According to Ackoff, social systems are purposeful systems with purposeful parts. The parts as well as the whole have the choice of both ends and means. Therefore, interaction between purposeful parts can take many forms, including conflict and cooperation. In conflict each party, by its presence, reduces the expected value of outcome for others. The opposite is true of cooperation. But it is by the following definition of competition that Russell Ackoff provides a key insight for understanding the concept of dialectic.

> **COMPETITION:** A and B are in competition when a lower-level conflict serves the attainment of a commonly held higher level objective for both $\underline{A}$ and $\underline{B}$.[19]

This leads us to the proposition that, in general, the purposeful actors (individually or in groups), by agreeing or disagreeing with each other on compatibility of their ends, means, or both can create the following four types of relationships:

1) *Cooperation:*   Compatibility of both *ENDS* and *MEANS*
2) *Competition:*   Compatibility of *ENDS*, incompatibility of *MEANS*
3) *Collaboration:* Incompatibility of *ENDS*, compatibility of *MEANS*
4) *Conflict:*   Incompatibility of both *ENDS* and *MEANS*

MEANS Compatibility	COLLABORATION	COOPERATION
Incompatibility	CONFLICT	COMPETITION
	Incompatibility	Compatibility
	ENDS	

Using this conception, dialectic seems to be similar to competition. It represents a conflict situation, in which contradictions in lower-level tendencies become the *means* for the creation of higher-level orders; a love-hate relationship.

Although each tendency tries to pull the system in its desired direction, the presence of the opposing tendency is also required to attain the system's end. The interdependence of the opposing tendencies are such that the seeds of the destruction of any system lies in the success (or the dominance) of one of the tendencies over the other.

In summary, *a dialectic* is a conflict in tendencies that share a higher-level objective. It is a conflict of *means* not *ends*. Whereas *a dichotomy* is conflict in the *ends* and *means*, a zero-sum-game and a win/lose struggle.

To the extent that members of a social system share a common image of a desired future, and agree on common ends, they are able to engage in mutually advantageous relationships. The interaction of opposing tendencies does not result in a state of frustration and immobility, which would happen if the tendencies were dichotomously related.

4.1 Changing a Dichotomy (win/lose) to a Dialectic (win/win) Environment.

One of the important characteristics of a win/lose struggle is the possibility of its conversion to either a lose/lose or a win/win environment. In the realities of present complex and highly differentiated social systems, the emergence of a lose/lose environment is not only highly probable, but it is an increasingly dominant reality.

Nowadays, winning requires much greater ability than ever before. It has become easier for any group to prevent others from winning than to win themselves. Increasing numbers of small special interest groups are diluting the strength of the traditional power centers. Even many disadvantaged minorities have been forced to learn how to prevent the opposing sides from winning. But the illusion that increasing losses for the other side is equivalent to winning is *the* reason for prolonging the struggle and playing the game to a lost/lose end.

Ironically, it is the awareness of this high probability for lose/lose that becomes instrumental in converting a win/lose struggle to that of a win/win. This is easily confirmed by understanding the reason why the players in the famous prisoners dilemma[20] chose the win/win strategy to avoid a lose/lose end. Here again the awareness of the possibility for lose/lose and the dynamic interaction of players creates a meta-game leading to a win/win.

On the other hand, ends and means are interchangeable concepts; an end is a means for a further end. Changing a dichotomous conflict to a dialectical one requires finding higher-level objectives shared by lower-level conflicting tendencies. This will transform the lower-level opposing ends to conflicting means that share a higher-level objective, thus a dialectic. The search for finding a shared higher-level end can continue up to and include the ideal, when ends and means converge and become the same.[21] The probability of finding a shared objective increases by moving to higher and higher levels, and is maximized at the ideal level.

Now, if even the ideal level cannot produce a common end for conflicting tendencies, then the conflict is considered nondissolvable within the context of existing world views.

In this situation, dissolving the conflicts requires a change of world views. This change can happen as a *reaction* to frustrations produced by failure of the existing assumptions to deal with the emerging realities of a new era, a march of events nullifying conventional wisdom. Or, it can happen by an active learning and unlearning process of purposeful transformation.

It is important to realize that more often than not the conflicting tendencies agree on the objective as a means to another conflicting higher end (collaboration). This is the cases in which a dichotomic conflict is converted to a dialectical one, only to be succeeded by another dichotomy at the higher level.

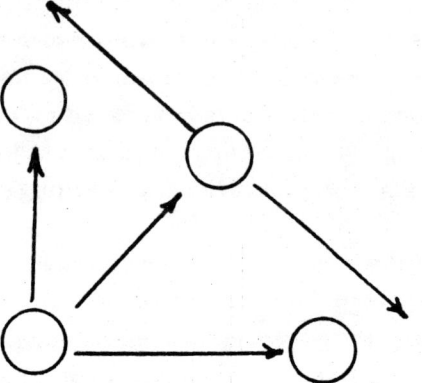

Figure 10. A dialectic leading to dichotomy.

To have a dialectic leading to another dialectic requires an agreement at the ideal level.

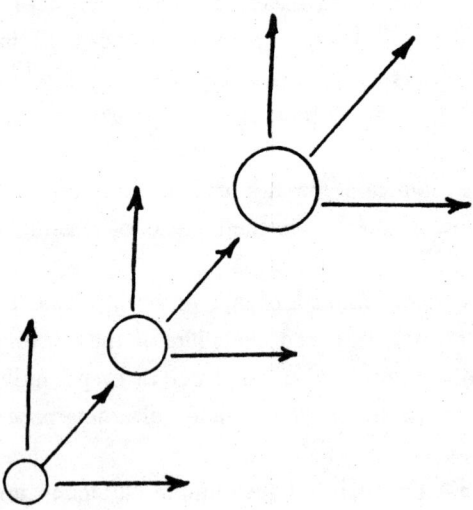

Figure 11. A dialectic leading to another dialectic.

But the ideal is not a fixed concept, an absolute. It is subject to constant reformulation by the members of the social-system reflecting their desires and fears of the *here* and *now*. Therefore, the agreement on the ideal also has to be reconfirmed accordingly. This is done not by agreement on an absolute ideal state, but on creation

SOCIAL DYNAMICS: DICHOTOMY OR DIALECTIC

of an ideal-seeking-system capable of purposeful transformation to higher levels.

SUMMARY

It is my contention that the ability of a social system to redesign its structure, redefine its function, and form an *AND* relationship between the opposing tendencies, creates a capacity for a purposeful and simultaneous transformation toward higher levels of differentiation and integration. It results in the possibility that a change of mode could take place without an observable cusp, or even a significant sign. This represents a new challenge; designers of a social-system must be alert to the possibility of a change of mode. If serious crises are to be avoided, critical assumptions regarding the nature of relationship among relevant variables must be repeatedly checked and verified. Constant verification is required even for those relationships that have been established experimentally over and over again, and are accepted as indisputable facts.

The courage to question the most sacred assumptions is a basic requirement for an active and continuous process of *learning and unlearning* which underlines the function of the control, guarantor, in a social system and the need for the design of an ideal-seeking-system.

Interactive Planning, developed by Russell Ackoff,[22] is a design methodology for creation of ideal-seeking-systems. It is particularly useful in the effort to change a win/lose struggle to a win/win strategy.

The five phases of an interactive planning process are (1) formulating the mess, (2) ends planning, (3) means planning, (4) resource planning, and (5) implementation and control.

In the phase of formulating the mess[23] — or defining the problematique — reference scenarios are used to project into the future an exaggerated version of existing tendencies to create awareness for a high probability of a lose/lose and to find the possible seeds of systems destruction. To the extent that the reference scenarios succeed in presenting a believable lose/lose future it can generate the required sense of crisis and the susceptibility to change. Finally in the ends planning phase the idealized redesign of the system not only provides a proper setting for a continuous learning process, but also the vision and motivation for redesigning the future and the means to bring it about.

The above principles have been successfully applied in more than 20 different cases of conflict situations. It is beyond the scope of this chapter to describe them here. The most recent was dissolution of a profound labor/management conflict which had originally led to closing of more than forty stores in the Philadelphia area. A win/win situation emerged with the formation of a new subsidiary.

NOTES

1. See R. Ackoff, "The Art and Science of Mess Management," Vol. 11, No. 1. *Interfaces*, 1981, and the *Art of Problem Solving*, John Wiley, 1978, page 13.

2. This is not an inclusive OR but an exclusive OR (XOR).

3. Complementarity is not a logical concept nor is the notion of the whole.

4. A coproducer, unlike a cause, which is both necessary and sufficient for an effect, is only necessary but not sufficient. For a full discussion of these concepts see R. Ackoff, *Scientific Method*, John Wiley, 1962, pages 16 and 311.

5. W. C. Churchman, *The Systems Approach and Its Enemies*, Basic Books, 1979, page xi.

6. C. Burrel, and G. Morgan, *Sociological Paradigms and Organizational Analysis*, Heineman, 1979.

7. Theories basically concerned with deep seated structural conflict and contradictions. (Burrel & Morgan, 1979)

8. Theories primarily concerned with stability and underlying unity and cohesiveness of social systems. (Burrel & Morgan, 1979)

9. G. Gordon, and Colleagues, "A Contengency Model for the Design of Problem Solving Research Program," *Milbank Memorial Fund Quarterly*, 1974, pages 184-220.

10. See W. C. Churchman, 1979.

11. E.C. Zeeman, *Catastrophe Theory: Selected Papers*, Addison Wesley, 1977.

12. See J. Gharajedaghi, "Organization as Information Bonded Systems," Industrial Management Institute, Publication 1972.

13. Blake and Mouton, *Managerial Grid*, Gulf Publishing Co.

14. For further discussion on dimensions of social systems see J. Gharajedaghi, "On the Nature and Management of Social Systems," S^3 *Papers*, 1981.

15. L. Althusser, "Overdetermination and Contradiction," No. 41, Jan., Feb., *New Left Review*, 1967.

16. K.E. Boulding, *Ecodynamics*, Sage Publications, 1978, pages 19 and 21.

17. W. Buckley, *Sociology and Modern Systems Theory*, Prentis Hall, 1967.

18. R. Ackoff and F. Emery, *On Purposeful Systems,*, Intersystems Publications, 1982.

19. Ibid.

20. See A. Rapoport, "Escape from Paradox," *Scientific American*, 1967 and A. Rapoport, and Ghammah, *Prisoner's Dilemma*, Ann Arbor, 1965.

21. R. Ackoff, in a discussion of the concept of omnicompetence argues that "one can desire nothing without desiring the ability to satisfy it. The ability to satisfy all desires is an ideal necessarily shared by all men at all times." This is a meta-ideal, where ends and means converge and become the same (1978, page 15).

22. R. Ackoff, *Creating the Corporate Future*, 1981.

23. E. Vergara, J. Gharajedaghi, and R. Ackoff, *Guide to Interactive Planning*, S^3 Papers, 1980.

4
ON THE NATURE OF DEVELOPMENT*

Development is a core concept of the systems view of the world. In contrast to the mechanistic and organismic views which are concerned respectively with *efficiency* and *growth*, the systems view is basically concerned with development.[1]

A critical review of major traditional views of development suggests that they are generally characterized by problems of (1) ethnocentrism, (2) unidimensionality, and, on the whole, (3) deterministic perspective.

In the first place, most developmental theories have built-in ethnocentric biases. The models, as ideal types of developed society, bear unmistakable signs of the western historical experience.

The second problem lies in their unidimensionality. The fragmentation of developmental theory into competing disciplinary perspectives has given us a fragmented view of development — preoccupation with material quantities in economics, power in political science, and order in sociology. The perspective of each discipline tends to exclude the other variables from its own unique domain of analysis.

The third, and perhaps the most serious, problem lies in their deterministic perspective. Because they begin with a preconceived law of social transformation — assumed to be true at all times and in all environments — the path is charted beforehand.

Misconceptions about the nature of development and the properties usually identified with it call for at least a clarification of the systems view of development and its relationship to other views.

In this chapter I will attempt to do this first by proposing a typology of major theoretical traditions that have contributed to our understanding of this complex process; and then deal with the telosystemic concept of development as I understand it.

---

*This paper originally appeared in Human Systems Management, Vol. 4, 1984, Elsevier Publishers B.V. (North-Holland).

1. A Typology of Major Theoretical Traditions in Development

A full discussion of various developmental theories is beyond the scope of the present work. However, the following typology provides a perspective for the introduction of the systems view of social development and its contrast to other major theoretical traditions.

Although it is risky to lump the great diversity of developmental theories together, for practical purposes we need some kind of classification scheme. This is despite the knowledge that it may obscure important differences and some significant continuities that exist among them. Further, it is important to note that these theories do not necessarily refute each other. In most cases they either complement or supercede one another.

In the typology presented below, developmental theories are categorized into eight types depending on their underlying assumptions (explicit or implicit) with regard to the singularity or plurality they attribute to function, structure and process.[2]

By *function* of a social system I mean outputs that it produces or the ends that it pursues. *Structure*, on the other hand, defines the components of the system and their relationships; and *process* deals with the mechanism of social transformation.

Singularity, refers to theories in which a particular structure, function or process is considered to be fixed or primary in all environments.

Plurality, refers to theories which consider structure, function or process to be multiple and/or variable in the same or different environments.

The eight categories of developmental theories are shown in Table (1). Letters F, S and P represent function, structure and process respectively. The small circle above each letter signifies singularity of that variable.

	Singularity of Function		Plurality of Function	
	Singularity of Process	Plurality of Process	Singularity of Process	Plurality of Process
Singularity of Structure	(1) F̊ S̊ P̊ Classical Neo-classical	(3) F̊ S̊ P Behaviorism	(5) F S̊ P̊ Structural/ Functionalism	(7) F S̊ P General Systems & Cybernetics
Plurality of Structure	(2) F̊ S P̊ Orthodox Marxism Radical Weberian	(4) F̊ S P Radical Humanism	(6) F S P̊ New – Left	(8) F S P Systems Views (Telosystemic)

Table (1) Typology of Developmental Theories

ON THE NATURE OF DEVELOPMENT

Note that theories in Category (1) (singularity of function, structure and process) are descriptive theories which do not deal with any means of intervention. Other categories by assuming plurality in at least one dimension provide for some means of intervention. Category (8) (Systems Views) assumes plurality in all three dimensions — function, structure, and process. Therefore, it is basically concerned with choice and purposeful behavior.

The following schematic summarizes the assumptions and the main features of each type and their perspectives on development.

***

A SCHEMATIC VIEW OF THE MAJOR THEORETICAL TRADITIONS

TYPE I	: Singularity of function, structure and process.
MODEL	: Predetermined, mechanistic and descriptive model of man in a state of nature, homo economicus, forms social contract to increase wealth through increasing productivity and division of labor.
THEORETICAL TRADITIONS	: *Classical and Neo-Classical* exemplified by writings of: Smith, Ricardo, Malthus, Mill, Marshall, Keynes, Schumpeter, and Rostow.
DEVELOPMENT PROCESS	: Stability and growth against major constraints of capital accumulation, population growth and limited natural resources; automatic mechanism of adjustment. Keynes introduces the principle of conscious manipulation of productive forces (neo-classical) to maintain stability and growth. Rostow considers a stage theory, traditional, pre-take-off, take-off, self-sustaining growth and high mass consumption.
TYPE II	: Singularity of function and process with plurality of structure.
MODEL	: Deterministic, mechanistic model based on linear cause and effect relationships. Conflict, the prime producer of change, results in a stage theory and formation of a new social structure.
THEORETICAL TRADITIONS	: *Orthodox Marxism, Radical Weberianism* exemplified by writings of: Engels, Lenin, Kautsky and Plekhanov; Weber, Dahrendorf and Rex.

DEVELOPMENT PROCESS	: In case of orthodox Marxism: economy is the prime function and class struggle is the prime process. Historical determinism, moving from primitive communism to ancient slave societies, feudalism, capitalism, socialism finally the ideal of communism (class-less-society) through class conflict and progressive system transformation.
	As for radical Weberianism, power is the prime *function* and legitimation is the prime *process*; varying structure defined by authority typologized into three pure types to correspond with different types of society: traditional charismatic and rational-legal. Increasing rationalization of authority from patriarchal to patrimonial to feudal and modern society moving toward ideal type of bureaucracy (friction-less-machine).
	Dahrendorf sees the interest of the power holder so clearly distinct from the interest of powerless that conflict becomes the permanent feature of social life, with varying degrees of effect, ranging from revolution to small-scale reform.

TYPE III	: Singularity of function and structure with plurality of process.
MODEL	: Input/output (stimulus-response) model of human and social behavior (environmentalism). An organic-machine, which learns through positive and negative feed-back (deviation amplification).
THEORETICAL TRADITIONS	: *Behavioral* exemplified by the writings of: Watson, Skinner, Erikson and Lasswell.
DEVELOPMENT PROCESS	: Increasing order through induced motivational and behavioral change. Sublimation of the destructive instincts into creative work; and finally formation of a world culture shaped by "behavioral technology," which is needed for survival. Watson places the central emphasis on controlling behavior through learning, which, he believes, could be achieved by the principle of "conditioning." Skinner suggests that freedom is an illusion which man can no longer afford. He claims that behavior can be predicted and shaped exactly as if it were a chemical reaction. But for Erikson, physical, social, cultural, ideational environments are partners to biological and psychological innate processes.

TYPE IV	: Singularity of function with plurality of structure and process.
MODEL	: There is no absolute above man, which could recreate the social order in which he lives. Emancipation of man is the prime

function; whereas, process and structure are seen as multiple and variable.

THEORETICAL TRADITIONS : *Radical Humanism* exemplified in the writings of: early Marx, Marcuse, Lukacs, Sartre, Fromm, Gramsci and the Frankfurt School.

DEVELOPMENT PROCESS : Changing the social order through a change in mode of cognition and consciousness. Release from the constraints which the existing social structure places upon human development. The emphasis is upon modes of domination, emancipation, deprivation and potentiality.

TYPE V : Singularity of structure and process with plurality of function.

MODEL : Organismic, integrated and dynamic equilibrium model, multiple functions to maintain an unstable but fixed structure (steady state) through the prime process of homeostasis. Analytical positivistic and empirical view of the world.

THEORETICAL TRADITIONS : *Structural-Functionalism* exemplified by the writings of: Comte, Spencer, Durkheim, Parsons and Eisenstadt.

DEVELOPMENT PROCESS : Integration, adaptation, goal attainment and pattern maintenance are regarded as the four functional imperatives of a social system for its continuing existence and evolution toward maturity and growth.

TYPE VI : Plurality of function and structure. Singularity of process.

MODEL : Multifunction, organic and non-linear cause and effect relationships. Conflict as the prime producer of change. Varying structure "overdetermined" by interaction of economic, political, ideological and theoretical sub-systems of totality.

THEORETICAL TRADITIONS : *New-left* exemplified in writings of: Althusser, Poulantzas, Della-Volpe and Colletti.

DEVELOPMENT PROCESS : Increased integration, through law of "uneven and combined development," "method of successive approximation," "fact of conquest," and increased accumulative knowledge of mankind with regard to nature.

TYPE VII	: Plurality of functions and processes with singularity of social structure.
MODEL	: Holistic, open, multi-loop feedback and input/output model of social systems. An organismic analogy, searching for the underlying regularities and structural uniformities.
THEORETICAL TRADITIONS	: *General Systems and Cybernetics*, exemplified by the writings of Bertalanffy, Ashby, Miller, Beer, and Bogdanov.
DEVELOPMENT PROCESS	: Equifinal, negentropic processes toward organized complexity. System change through learning, adaptation, and induced motivational and behavioral change.

TYPE VIII	: Plurality of structure, function and process. Multiple and variable concepts of structure function and process.
MODEL	: Purposeful, knowledge-bonded-system. Capable of structure, creation and maintenance.
THEORETICAL TRADITIONS	: Systems view (telosystemic) exemplified by writings of: Ackoff, Boulding, Buckley, Churchman, and Rapoport.
DEVELOPMENT PROCESS	: Multifinal, interactive, purposeful movement toward increased differentiation and integration. A learning and creative process to increase ability and desire to recreate the future. An ideal-seeking mode of organization to dissolve conflicts at higher levels.

---

2. *System's View of Development*

The development of a social system is a learning and creative process by which a social system increases its **ability** and **desire** to serve its members and its environment by the constant pursuit of: truth, plenty, good, beauty and liberty.[3] It results in a purposeful transformation toward increased integration and differentiation at the same time.

The two major components of development, therefore, are desire and ability.

Desire is produced by a vision enlarged through the interaction of creative and recreative processes.

The creative capacity of man, along with his desire to share, results in a shared image of a desired future. This generates dissatisfaction with the present and motivates pursuit of more challenging and more desirable ends. Otherwise, life proceeds simply with setting and seeking attainable goals which rarely escape the limits of the familiar.

Dissatisfaction with the present, although a necessary condition for change, is not sufficient by itself to ensure development. What seems to be necessary, in addition, is a faith in one's ability to partially control the march of events. Those who are awed by their environment and locate the shaping forces of their future only outside of themselves do not think of voluntary or conscious change, no matter how miserable and frustrated they are.

Ability, therefore, is the potential means of controlling, influencing and appreciating the parameters which effects the system's existence.

But ability alone cannot assure development. In the absence of a shared image of a more desirable future the frustration of the powerful masses can easily be converted to a unifying agent of change – hatred – which, in turn, will result in the successful destruction of the present but will not necessarily be a step toward the creation of a better future. The recent Iranian case is a good example. In most of the Middle Eastern countries a certain interpretation of Islam – the Fundamentalist – regards creation as sole prerogative of God. Human beings are assumed not to be capable of, therefore not allowed to engage in, any act of creation. Art in almost any form – painting, sculpture, music, drama – is prohibited. Recreation is also considered sinful.

This antagonistic attitude toward aesthetics militates against development, in that it does not provide much opportunity to articulate and expand one's horizon beyond the immediate needs of mere existence. This provides one explanation for cases of underdevelopment despite the availability of vast resources.

Central to systemic notion of development is its distinction from growth. According to Ackoff, "They are not the same thing and are not even necessarily associated. Growth can take place with or without development, and development can take place with or without growth. A cemetery can grow without developing. On the other hand, a person may continue to develop long after he or she has stopped growing, and vice versa."

"Growth, strictly speaking, is an increase in size or number. Its principal but not exclusive domain of relevance is biological, as in growth of plants and animals. Social groups are also said to grow when they increase in size. It would be nonsensical to speak of a growing culture because size and number are not relevant to it. An organization or a nation, like an individual, can grow by increasing in size or, unlike an individual, in number without developing; it can also develop without growing."[4]

Constraints on a system's growth are found primarily in its environment; but the principal constraints on its development are found within the system itself. Therefore the principal limits to growth are external; those to development are internal.

In this context, development is a potentiality for the satisfaction of desires, not the quality of life nor the standard of living actually achieved. The quality of life that a system can realize is the joint product of its development and the resources available to it. Although this implies that limited resources may limit improvement in the quality of life, it does not imply that they limit development.

As Ackoff puts it, "A man can build a better house with good tools and materials than he can without them. On the other hand, a developed man can build a better house with whatever tools and materials he has than a less-developed man with the same resources. Put another way: a developed man with limited resources is likely to be able to improve his quality of life and that of others more than a less-developed man with unlimited resources."[4]

Development of social systems does not follow the physical pattern, which is conceived to be a unidimensional movement toward increased complexity in the structure of the matter. And it is unlike biological evolution, which reflects a two-dimensional negentropic movement toward complexity and order. It is conceived to be a multidimensional and purposeful transformation into successive modes of organization. Each mode is a whole characterized by higher degrees of both integration and differentiation, and is potentially capable of dissolving lower level contradictions by converting them into contraries.[5]

In contrast to physical systems, in which the energy level determines their mode of organization, it is the knowledge level which defines it for social systems. Therefore, the role of knowledge in social systems can be said to be analogous to that of energy in physical systems. Although the change of phase in physical systems (solid-liquid-gas) is said to be analogous to the change in mode of organization in the social systems, (feudalism, capitalism and socialism) the analogy ends there.

The significant point is that knowledge, unlike energy, is not subject to the first law of thermodynamics (the law of conservation of energy). One does not lose knowledge by sharing it with others. On the contrary, its dissemination increases the knowledge level of the social system.

The capability of creating knowledge can be shown to enable a social system to constantly recreate its structure and redefine its functions. It makes possible a change of mode in social systems without a significant sign or a catastropic cusp.[6]

Development, as the process of creating successive modes of organization should involve (1) an active process of the generation and dissemination of knowledge, (2) a process of learning and adaptation, (3) the creative process of discovery of new dimensions with all of their implications, and, finally, (4) the painful process of reconceptualization, reformulation and integration of all the variables involved in a new ensemble with entirely new relationships and characteristics of its own.

ON THE NATURE OF DEVELOPMENT

Finally, the telosystemic view of development, by accepting plurality in all three dimensions of function, structure and process considers the other seven categories as special cases. From the systems perspective, development is not only a multifunctional phenomenon, but involves multiple and varying concepts of *structure* and *process* as well. This point requires further clarification.

(1) *Plurality of functions.* Historically, the identification of functional areas of social systems has been at least as reactive — reacting to certain problems in social life — as proactive — reaching for the ultimate good. However, it is interesting to note that although some prominent social thinkers have implicitly considered more than one dimension in their analysis, each one has, somehow, a chosen a *single* and not surprisingly different function as *prime cause* of all social phenomena. Marx for example considered the *economy*, the mode of production as the underlying cause of social realities. Whereas, for Weber *power*, supported by notions of authority and legitimacy seemed to be the prime concern.

Reactively the five dimensions of social systems corresponds to the following major problem areas historically faced by all human societies:

Economics — The generation and distribution of wealth, that is the production of necessary goods and services and their equitable distribution.

Scientific — The generation and dissemination of information, knowledge and understanding.

Aesthetics — The creation and dissemination of beauty, the meaningfulness and excitement of what is done in and of itself and the enjoyment derived therefrom.

Ethics — Creation and maintenance of peace, conflict resolution, the challenge of appreciating plurality of value systems.

Politics — The generation and distribution of power, questions of legitimacy, authority and responsibility — or, in general, the question of governance.

On the proactive side, Russell Ackoff identifies the same five dimensions in discussion of ideal seeking systems. Referring to the ancient Greek philosophers, he identifies four classes of societal activity that are individually necessary and collectively sufficient for progress towards the ideal, Omnicompetence."There are the pursuits of *truth* (scientific and technological function), *plenty* (the economic function), *good* (ethical-moral function), and *beauty* (aesthetic function). He concludes that, "to carry out these functions society must be organized and managed effectively. The way society is organized and managed are matters of *politics*" This is the dimension that in the definition of development given at the beginning of this section I added as the ideal of liberty.

The purpose of the above classification, unlike those of conventional practice, is *not* to isolate each dimension so it can be analyzed separately. On the contrary it is to emphasize the interactive nature of these dimensions. It excludes the concept of a "single leading factor" which for most developmental theories seems always to be in the forefront.

Nevertheless, because of the interdependence and mutual effect of the dimensions on one another it is quite feasible to use any one of them to explain the characteristics of other dimensions in part. But this is at best an over simplification, and underestimation of the complexity and immense potentialities of social systems.

(2) *Plurality of structure.* Earlier, we proposed that structure of a system defines the components and their relationship. Plurality of the structure, therefore, means that components and relationships among them are multiple and variable. Consider, for example, a substance, salt (NaCl). Its components — Chlorine (Cl) and Sodium (Na) — form a single type of relationship in all environments; therefore, salt is said to have a singular structure. But the same cannot be said about Hydro-Carbons. Hydrogen and carbon enter into various combinations and relationships resulting in multiple structures. In social context still more complexities are encountered. Aside from multiple relationships that can exist between components of a social system, the nature of the component itself is variable — individual members change through learning.

A social system is a purposeful system with purposeful parts. The parts as well as the whole have the choice of both ends and means. Interaction between these purposeful parts at any given time can take many forms. Actors (individually or in groups) by agreeing or disagreeing with each other on the compatibility of their ends, means, or both can form the following four types of relationships.

(1) *Cooperation:*   Compatibility of both *ENDS* and *MEANS*
(2) *Competition:*   Compatibility of *ENDS*, incompatibility of *MEANS*
(3) *Collaboration:* Incompatibility of *ENDS*, compatibility of *MEANS*
(4) *Conflict:*   Incompatibility of both *ENDS* and *MEANS*

These relationships may even coexist simultaneously. Actors in a social system may cooperate with regard to one pair of tendencies, compete over others and be in conflict with respect to different sets at the same time. The result is a dynamic concept of social structure; an interactive network of varying components with multiple relationships. Acceptance of plurality of structure, unlike that of functions, is a difficult proposition since it goes against a long standing traditional conception of structure as something which endures. However, a reconceptualization of this traditional conception of structure is required for understanding the telosystemic principle of purposefulness, that is the ability of a social system to redefine its functions and redesign its structures.

(3) *Plurality of Process.* The classical principle of causality maintained that similar conditions produce similar results, and consequently dissimilar results are due

ON THE NATURE OF DEVELOPMENT

to dissimilar conditions. Therefore for a given structure, behavior of the system is completely predictable and its future states invariably depend on its initial conditions and the laws which govern its motion (determinism). Bertalanffy, in analyzing the self-regulating or morphostatic features of open biological systems loosened this classical belief by introducing the concept of "equifinality:" a final state may be reached by any number of different developmental routes.[7]

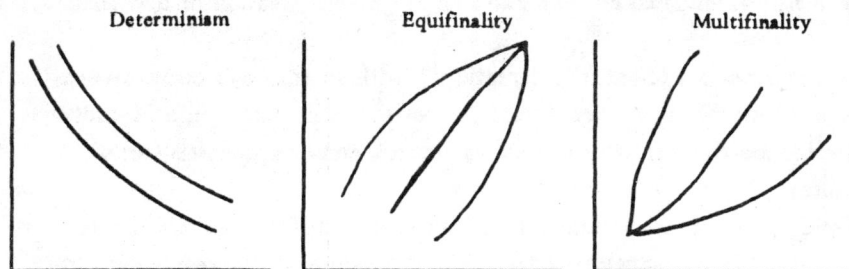

Walter Buckley in his discussion of morphogenetic processes in socio-cultural systems goes even further to suggest an opposite principle called "multifinality." Similar initial conditions may lead to dissimilar end states. So the process rather than the initial conditions is responsible for future states.[8]

Finally the concept of producer-product developed by E. A. Singer,[9] coupled with the notion of non-linear feedback loops (positive/negative) results in a network model of social causality where the cause and effect displace one another successively and mutually affect and are affected by one another.

The dynamics of the social system and the principle of multifinality can be understood by the notion that the sets of opposing tendencies, which are usually treated as dichotomies, are in fact two sides of the same coin. They coexist and interact continuously; so that the relationship between opposing pairs might be characterized by an "AND" rather than an "OR" relationship. For example, tendency toward security and freedom complement one another.

Freedom is not possible without security and security makes no sense without freedom; nevertheless, both might be achieved by a process called participation. Similar arguments can be made for other opposing pairs such as stability and change. Despite seemingly contradictory requirements for the pursuit of the opposing ends within a pair, they form a complementarity and coproduce a process which makes the attainment of both ends feasible. For instance, pursuit of both stability and change might be attainable by *adaptation*, that of order and complexity by *organization*, and uniformity and uniqueness by *innovation*.

Furthermore, it is interesting to note that security, stability, order, uniformity . . . seem to share a certain characteristic and belong to a set which can be termed *integration*; while freedom, change, complexity, uniqueness . . . manifest an opposing characteristic and belong to another set called *differentiation*. To generalize:

Differentiation represents an artistic orientation[10] with an emphasis on stylistic value systems, signifying tendencies toward: increased complexity, increased variety, increased individuality (individual choice), and morphogenesis (creation of new structure).

Integration represents a scientific orientation[11] with an emphasis on instrumental value systems, signifying tendencies toward: increased order, increased uniformity and conformity, increased collectivity (collective choice), and morphostasis (maintenance of structure).

The emerging processes, coproduced by interaction of differentiative and integrative tendencies, such as participation, adaptation, innovation, and organization cannot stand alone; together they form the whole, and coproduce the process called development.

In summary, plurality of structure, function, and process, manifested in the ability of a social system to redesign its structure, redefine its functions and exhibit a purposeful behavior, puts the social system in a class by itself. The class is of such a level of complexity that existing analogies — mechanistic or organismic — do not provide a relaistic model for its understanding.

The systemic view of societal development suggests that all of the five social functions — generation and dissemination of knowledge, power, wealth, value and beauty — develop interdependently utilizing the whole set of complementary integrative and differentiative processes to form an ideal-seeking mode of organization.

This concept of development is based on a socio-cultural model and the notion of participative democracy. It is a multi-dimensional phenomenon defined in such a way as to avoid ethnocentric and deterministic biases. Each social system is allowed to set its own course in terms of its perceived desires, whereby the uniqueness of the system's identity and culture are enhanced as development proceeds.

Practical implications of this conceptual framework are explored in a separate work (Organizational Implications of Systems Thinking) by considering a methodology for the design of social organizations as ideal-seeking-systems.[12]

ON THE NATURE OF DEVELOPMENT

NOTES

1. See Gharajedaghi, J., "The Why Question-Worldviews," S^3 Papers, 1981.

2. For the argument why structure, function and process together provide a holistic view of a system, see J. Gharajedaghi, "Systems View of Social Systems," S^3 Papers, 1982. In addition note that there are four attributes for each of the dimensions of *structure*, *function* and *process* based on whether they are seen as (1) fixed or variable, (2) single or multiple. This leads to $4^3=64$ categories.

	Fixed	Variable
Multiple	Fixed & Multiple	Variable & Multiple
Single	Fixed & Single	Variable & Single
	Fixed	Variable

3. This is a modified version of the definition of development proposed by Ackoff. See *Art of Problem Solving*, Wiley and Sons, 1978.

4. See Ackoff, R., and et al, *National Development Planning*, Busch Center, 1982.

5. Gharajedaghi, J., "Social Dynamics, Dichotomy or Dialectic," *Human Systems Management*, 1983.

6. Ibid.

7. Bertalanffy, L., *General Systems Theory*, Penguin Books, 1960.

8. Buckley, W., *Sociology and Modern Systems Theory*, Prentice-Hall, 1967.

9. Singer, E.A., Jr., *Experience and Reflection* . C. W. Churchman (ed), University of Pennsylvania Press, 1959.

10. Searching for differences among things which seems to be similar.

11. Searching for similarities among things which seems to be different.

12. Gharajedaghi, J., "organization Implication of Systems Thinking," *European Journal of Operational Research*, Vol. 18, No. 2, 1984.

5
OBSTRUCTIONS TO DEVELOPMENT*

We defined development as a purposeful transformation, learning and creative process by which a given system increases its desire and ability to serve its members and its environment by constant pursuit of: beauty, plenty, good, truth and liberty.

To understand the obstructions to the development of a social system we have to deal with structures and the processes which help or limit the creation of *collective desire* and *ability* in a social system for the pursuit of its ends.

There are a number of alternate ways of looking at obstructions. However, based on the above devinition, obstructions to development can be viewed as malfunctioning in any one of the five dimensions of the social systems. *Scarcity*, *maldistribution* and *insecurity* in any one of the five social functions, that is, *scientific*, *political*, *economic*, *ethical* and *aesthetic* are considered to be primary obstructions. Corruption, alienation, polarization and so on are among social phenomena that represent secondary obstructions to development. Secondary obstructions are coproduced by the interaction of more than one dimension. Therefore, their resolution requires some kind of structural change. (See the following table.)

A complete treatment of the first order obstructions can be found in *National Development Planning*.[1] In this chapter, three major second order obstructions — alienation, polarization and corruption — are discussed. "Social pathology" then, is, as a mode of organization incapable of removing a persistent obstruction to system's development.

1. Alienation

A social system, in its ideal form, is a voluntary association of purposeful members, so that emigration of a member from the system is considered to be the highest manifes-

---

*This paper originally appeared in Human Systems Management, Vol. 4, 1984 Elsevier Publishers B.V. (North-Holland).

Dimensions of Social Systems	Expected yield	Primary Obstructions (First Order)			Secondary Obstruction (Second order)
		State of scarcity	State of maldistribution	State of insecurity	
Economic	goods/services (Plenty)	Poverty inefficiency	disparity exploitation	fear of deprivation instability	A L I E N A T I O N P O L A R I Z A T I O N C O R R U P T I O N
Scientific	information knowledge, understanding (Truth)	Ignorance Incompetence Rolelessness	Elitism/illiteracy Lack of communication	Obsolescence	
Political	influence (participation) (Liberty)	impotency powerlessness	Centralization Autocracy	Illegitimacy	
Ethical/Moral	Peace (Good)	normlessness	Conflict discrimination	Fanaticism	
Aesthetic	Sense of belonging Excitement (Beauty)	meaninglessness hopelessness Boredom	Lack of shared Image of Desired future Selfishness/Selflessness	Fear of loss of identity & and individuality Fear of loneliness & isolation	

tation of his protest. However, because of a series of self-imposed or external constraints, a dissatisfied member may not be able or willing to leave the system. He, therefore, becomes alienated from the very system of which he is supposed to be a voluntary member. The underlying causes of *alienation* can be found in one or a combination of the following factors, each corresponding to one of the five dimensions of the social system.

Powerlessness. Powerlessness is equivalent to ineffectualness and impotency. When an individual feels that his contributions to the group's achievements are insignificant, or when he feels powerless to play an effective role in the system's performance, gradually a feeling of indifference sets in and he becomes alienated from the very systems of which he is supposed to be a part. The feeling of powerlessness is due in part to the organizational set-up which is usually designed mechanistically, thus forcing a passive functioning of the parts. Furthermore, just as the strength of a chain is determined by its weakest link, so too, incompatibility between the strength of various elements in a developing system often causes the more dynamic units to retrogress to the level of the weakest, spreading a general feeling of ineffectualness and impotency (Power dimension).

Rolelessness. Tangible meaning and significance of purposeful information-bonded systems (family, group, organization, nation) lie in the fact that the unit of these systems is not so much the individual but the role imparted to him/her. Under different sets of circumstances and in different social settings individuals display different behavior. A good friend is not necessaily a good employee, a successful vice-president might make a lousy president. The nature of these roles is influenced by expectations and limitations imposed by the social structure, the culture, and various environmental realities mapped by individuals.

Lack of sufficient knowledge and proper professional skills to carry out responsibilities of a specific role (Incompetence) result in excessive anxiety and frustration. To fulfill the role of a physician or carpenter requires certain expertise and mastery which must be learned, otherwise, the individual to whom the role is entrusted will be alienated (Knowledge dimension).

Meaninglessness. Lack of a meaningful, exciting and challenging mission in life, suppression of an individual's need for creativity and achievement, and finally insensitivity toward the recreational aspect of the production process are probably the main causes of meaninglessness. An individual whose only duty consists of writing entries in a book, tightening a bolt, or placing stickers on a finished product cannot find sufficient meaning in the activity to satisfy his need for achievement and creativity. He cannot relate his contributions to the totality of the product produced or service rendered. Then, the work, instead of fulfilling a set of needs and desires, becomes a means of fulfilling material needs alone, alienating the individual from the very act of production itself (Aesthetic dimension).

Exploitation. The feeling of injustice in the fair distribution of a system's achievements is another factor which can cause alienation. When an individual feels that he has somehow been deprived of his fair share of recognition for contributing to a system's achievements, he becomes alienated and frustration will result (Wealth dimension).

Conflicting value system. Finally, conflicting values within a social system contribute to alienation of its members. As mentioned before, the extent of which an individual's value image coincides with the "shared image" of his community, determines the degree of his membership. The level of integration that a society will achieve depends on the means by which it dissolves the conflict among its members. A certain degree of consensus with regard to desired ends is required for continued and productive membership in a given system.

2. *Polarization*

Polarization of a population around conflicting ideologies is another important secondary obstruction witnessed in almost all developing nations. The formation of highly polarized political and social groups is the most destructive phenomenon confronting the majority of underdeveloped nations. This polarization is further reinforced by ethnic conflicts and the "divide and rule" strategies of local and foreign powers.

In most Middle Eastern countries polarization takes the form of religious versus secular tendencies, with each of these further divided into groups of leftist and rightist, with national or international orientations.

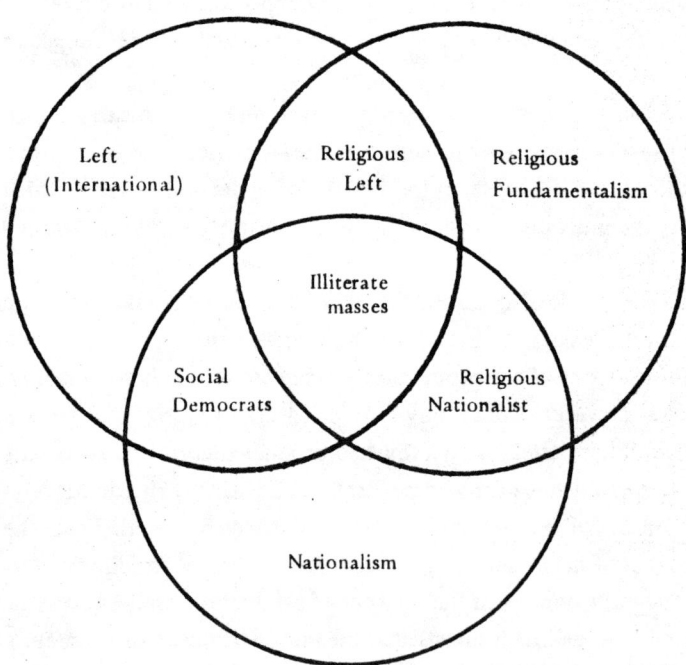

The problem is that no one of the so-called opposing groups is strong enough to govern without the cooperation of the others, and yet each one is powerful enough to disrupt and undermine the effectiveness of the ruling group. This is partly due to increased complexity in the system, making it more vulnerable to sabotage on one hand, and more difficult to manage on the other.

OBSTRUCTIONS TO DEVELOPMENT

Once it seizes power, the ruling group sets out to monopolize the power. This brings opposing groups together in a collaborative effort to paralyze the government, which in turn intensifies its efforts to eliminate the opposition through oppressive means. Hatred of the ruling group thus becomes the unifying agent, and the opposition movement erupts and results in destruction. New leaders which emerge soon renew the oppressive measures of the previous regime. Hatred grows anew, another cycle begins with opposing forces regrouping. Ironically any conciliatory gesture by the oppressive regime towards the opposition is taken as a sign of weakness. This causes them to intensify rather than modify their attacks on the government. The regime is forced to tighten its control over the opposition until neither more pressure nor more freedom can save it from collapse.

Interestingly enough, this process does not depend on the ideology of the ruling group. There is no example of one which has been able to break out of this vicious circle. The secular left in Iraq and Afghanistan, the religious left in Libya, and both the religious and secular right in Iran and Pakistan provide new supportive evidence for this argument.

Even in the countries where a change of power takes place less abruptly, for instance Turkey, polarization has resulted in increased terrorism, complete paralyzation of government, and oscillation between extremes. The major effort of every new government has always been to undo every change instituted by the preceding regime, be it industrial projects or elementary school text books. The new regimes in Iran and Turkey have both set out to rewrite school books. Even the cultural heritage is not spared. Every new regime disbands existing historical accounts calling them false and misguiding, and presents it own version as correct. The plan is always to destroy the old system completely before starting to construct a new order. Not surprisingly, the first part is usually more successful than the second, breaking the cumulative cultural and developmental continuity between one period and the next.

Self-serving education elite and the great number of illiterate masses is another factor contributing to propagation of polarization. Self appointed guardians of the working class and cynical intellectuals, in their struggle for power and fame, manipulate the masses with slogans and demagoguery pulling them from one extreme to the next like a pendulum.

The oscillation will not end until opposing groups learn to modify their dogmatic positions, give up their monopolisitc claim on power, and work towards creating a shared image and consensus among people through processes of integration, not at the expense of differentiation, but alongside it.

We have suggested that development is a creative learning process through which the collective desire for and ability to pursue a more desirable future is increased. Illiteracy is among the primary obstructions which impede this process in at least the following respects.

Social systems, in view of their information-bondedness, must provde appropriate communication channels for exchange of information among their members if they are not to disintegrate into aggregrates of individuals. Communication channels play an essential role in the creation of a common image, as well as in dissemination of variety and mapping process within a system. The written language is a very important channel since it is delimited neither by the spatio-temporal constraints of direct communication, nor by the techno/economic constraints of tele-communication. It is the medium through which history and the cultural heritage are best preserved and manifested in the common image.

Written communication provides a much better opportunity to reflect and to learn than does spoken language. A book can be picked up at any time, read through at any speed, and re-read as many times as desired. The message is not lost if it is missed the first time. Control is mostly at the receiver's end rather than the sender's. The reader enjoys freedom in terms of "what" and "why" can also decide "when" and "how." Finally, familiarity with a new symbolic system, such as written language, can potentially increase the power of creativity through exercises in symbolic manipulation.

3. Corruption

Corruption is a multi-dimensional phenomenon. It is not just a malfunctioning of the value system, but a second order obstruction whose influence extends to all dimensions of social systems, including the generation and distribution of Power, Wealth and Knowledge. Corruption is an inevitable consequence of machine age management philosophy, which is not only foreign to complex cultures but, in view of the increasing complexity and accelerating rate of change, has lost its effectiveness even where it was originally conceived. It was developed to manage things, not complexities, by dealing with a single problem at a time.

At the present level of interdependence and complexity, developing nations do not have the luxury of dealing with their developmental problems in progressive stages as did the more developed nations. Less developed countries have to solve their production problems while facing increasingly pressing problems of distribution. They have to solve their marketing problems in view of a shorter product-life cycle, economies of scale and challenge of obsolescence. They have to deal with sophisticated technology, and its increasing requirement for specialization, with an army of unskilled, impatient, demanding labor, and inexperienced, mistrained and incapable technocrats. They have to create social stability under the overwhelming pressure of opposing and antagonistic idealogies.

Ironically, the necessity for coping simultaneously with interacting sets of problems represents the most important opportunity for a less developed nation — the opportunity to avoid unforeseen consequences of sub-optimization and to redesign the future with all its dimensions in mind. However, to realize this oppostunity requires a radical change in management philosophy. This, in turn, demands a higher degree of awareness and sophistication which unfortunately is not yet in sight. On the contrary the challenge of meeting the primary requisites of economic development, sufficient generation of goods and services, and their equitable distribution has created a growing tendency towards a closed, bureaucratic system.

Unfortunately, bureaucracy leads inevitably to corruption, inefficiency, and resistance to change. Handling an interrelated network of problems requires an ability which far surpasses that of a bureaucratic system. A system in which people are promoted to their levels of incompetence (Peter's Principle) where they finally precipitate and resistance to change becomes infinite.

Under these conditions, only a source of power from outside the bureaucracy can create movement within the system. Therefore, individuals will seek out and support these external power sources which derive their influence and authority from higher levels. This leads finally to a strong man in charge, a paternalistic solution. In time, the hierarchy of powerful informal patrons will demand certain rewards in exchange for their valuable support. This allows corruption to spread throughout the entire system, ultimately to become a justifiable way of life.

A social system in any particular stage of its development, may require certain unusual strengths to survive and function acceptably in a turbulent and aggressive environment. Unfortunately, the strong features of a system often are associated with certain indispensable weak points. Eliminating the weak points without comprehending their interrelations with strong aspects, or concentrating exclusively on the strong features of a system without minding the accompanying weak points will lead to the overall malfunctioning of the whole system in the long run.

Accordingly the market economy and centralized planning, at opposite ends of a continuous spectrum, have both strong and weak features. Therefore, a random mixture of the two, which is usually employed by developing countries, becomes doubly ineffective since it may incorporate only the weak features of both.

The main advantage of the market economy is in its production power, which stems from competition. Its disadvantage is in the distribution system with its natural tendency toward monopolies. The allocation of resources becomes more wasteful and less equitable especially in the absence of equitable distribution of wealth, due to mechanisms which register demand only through dollar-vote and purchasing power.

In a classical market economy, the mechanisms of supply and demand determine the type, quantity, and quality of outputs. The "pricing of goods/services" is supposed to be reliable and sufficient criterion for determining production priorities. This supposition might be tenable if end-prices were not manipulated. However, factors wuch as price control or government protection make the actual cost of the goods and services much higher than perceived. In other words, inputs are purchased from the environment at a lower price and outputs (measured by the conventional accounting method) appear more valuable than they really are. In the absence of equitable distribution of wealth the needs of the majority have little influence in determining the actual demand. In most cases it is the dollar-vote not the individual vote that registers the demand. For instance, suppose that in a society of fifty, ten have a purchasing power of $100,000 each and the other forty $5,000 each. Then the questions of "what to produce?", "how much to produce?" and "for whom to produce?" are answered more by the needs and wants of the ten rather than the forty; and the economy moves towards the production of goods to satisfy the wealthier echelon of society. Moreover, protecting a particular production activity and ignoring the relative advantages of other economic opportunities increases prices in the following manner. Economic activities with unusually high rates of return on investment will attract limited resources at a higher cost. The higher cost will be transferred to consumers who are not very sensitive to price increases. This maintains the rate of return on investment, and the vicious circle continues.

The strong points of centralized planning are perceived to be the effective allocation of resources and equitable distribution of outputs. Its main problem lies with production and bureaucratic orientation. Bureaucrats and technocrats tend to monopolize the right to issue permits, to the degree that they insist upon permits for the pettiest matter possible, claiming this promotes coordination and standardization. The corruptible bureaucrat, by granting monopoly rights to certain individuals, enters into a second-degree agreement with the corrupting entrepreneur who is willing to wipe out competition at any cost. With competition destroyed, quality degenerates rapidly and administrative corruption completely paralyzes the distribution system which was supposed to be the strong feature of centralized planning.

Apart from bureaucratic corruption, lack of proper motivation aggravates the already severe problems of production. The inherent inability of the bureaucratic system to generate required goods and services results in nothing but equitable distribution of poverty. This is not to under-value the importance of an equitable distribution system but to emphasize its inseparable connection with production. As mentioned before, without an effective production system there can never be a meaningful distribution system. Production and distribution may be viewed and discussed separately, but they cannot be separated. Failure to note this significant interrelationship is to leave out the most important challenge of the problem. Therefore, obsession with distribution without a proper concern about production, (nowadays a powerful political tool and fashionable demagoguery) will result in nothing but further frustration.

Ironically, in most of the developing countries it seems as though the government and its bureaucracy has taken over the responsibility for production leaving the private sector and market economy in charge of distribution. Therefore, it is not difficult to realize why everyone, especially bureaucrats, are engaged in an intense game of speculative middlemanship.

In sum, by taking into account both the strong and weak points of the market economy and centralized planning systems, we realize that the ultimate answer lies in neither one nor in a mixture of the two. It must be sought in a creative synthesis on a higher level. We will face this challenge later in the last section of the present work.

SOCIAL PATHOLOGY

In order to carry out its functions and to deal with obstructions to its development, social systems must be organized and managed effectively. The way a social system is organized stands at the center of its change process. This brings us to the concept of organization and its patholoty.

The word "pathology," in a biological context, is defined more or less as deviation from the norm. However, in the social context, it has taken an opposite meaning. It refers to the inability of a social system to change itself. Social pathology is defined as:

> "The inability or lack of desire in a *government/management* of a social system to remove a persistent obstruction to development that can be removed by a change in either (a) the way the system *is organized* and/or (b) its social or physical environment."[3]

A pathology is produced when an obstruction to development benefits some who have the ability and desire to obstruct its removal. Therefore, bureaucracy, technocracy, theocracy, aristocracy . . . each represents a pathological mode of organization. In each an organized interest group dominates the management or governance of the social system.

Traditionally management has dealt with organizational concepts utilizing an implicit model of the organization which is a replica of mechanistic or organismic systems.

The mechanistic models view the social system as a mechanical device that tends toward a predetermined equilibrium point. Thus given a certain structure, the behavior of the system is completely predictable. The machine is the metaphor of such views of social systems.

The organismic models, by contrast, employ the organismic metaphor for the understanding of society. The social system is viewed as a structure-maintaining organism that unfolds according to a genetic "blue print" and moves toward essentially the same ultimate destination. The principle of equifinality[4] is at work here.

Socio-cultural models of social systems emerging out of systems thinking challenge the assumptions of both the mechanistic and organismic models. Social systems are viewed as information/culture-bonded systems engaging in structure elaboration as well as structure creation. Its developmental path is neither predetermined mechanistically nor is organismically fixed in destination — rather it is multifinal[5] and purposeful.[6]

The major differences between mechanical and social systems are so pronounced that those principles which make the design and control of a mechanical system so successful cannot be used to meet the challenges of managing complex social organizations. For a discussion of the systems view of organization see the "Organizational Implications of Systems Thinking."[7]

NOTES AND REFERENCES

1. Ackoff, R. and Gharajedaghi, J., *National Development Planning*, Busch Center, The Wharton School, University of Pennsylvania, 1983.

2. Boulding, K., *The Image*, Ann Arbor Paperbacks, The University of Michigan Press, 1956.

3. This definition of social pathology was developed in a learning cell at the Department of Social Systems Sciences, The Wharton School, along with Professors R. Ackoff and A. Katsenelinboigen.

4. Bertalanffy, L., *Problems of Life*, Harper and Row, 1960.

5. Buckley, W., *Sociology and Modern Systems Theory*, Prentice-Hall, 1967.
6.
6. Ackoff, R., and Emery, F., *On Purposeful Systems*, Intersystems Publications, 1982.

7. Gharajedaghi, Jamshid, "Organizational Implications of Systems Thinking," *European Journal of Operational Research*, 1984.

6
ORGANIZATIONAL IMPLICATIONS OF SYSTEMS THINKING*

We have argued extensively elsewhere[1] that the development of social systems is the process of creating successive modes of organization at higher levels of complexity and order. We have gone so far as to define *Social Pathology* as a mode of organization incapable of changing itself. Therefore, the ability and desire to change, as manifested in learning and adaptive systems, becomes the core concept of the systems view of organization. Now we have to propose a model of organization, which not only overcomes the shortcomings of mechanistic and organismic equilibrium models, but also results in a participative mode of organization whose function is to serve the purposes of its purposeful members as well as its purposeful environment.

We will face this challenge by introducing the concept of multidimensional modular design. But first we need to review the evolution of organization theories. The concept of multidimensional modular design has evolved out of fifteen years of real life experimentation with organizations in different cultures. However, although variations of this design have been successfully implemented in more than fifty organizations,[2] its unconventionality makes it difficult for some conventional managers to accept.

1. EVOLUTION OF ORGANIZATION THEORIES

Traditionally, organizational structure is understood to deal with two types of relationships: (1) responsibility — who is responsible for what, and (2) authority — who reports to whom. Structure, so conceived, lends itself to representation by a two-dimensional chart in which boxes represent responsibilities and levels and lines represent the loci and flow of authority.

---

*This paper originally appeared in European Journal of Operational Research, Vol. 18, No. 2, 1984. Elsevier Publishers B.V. (North-Holland)

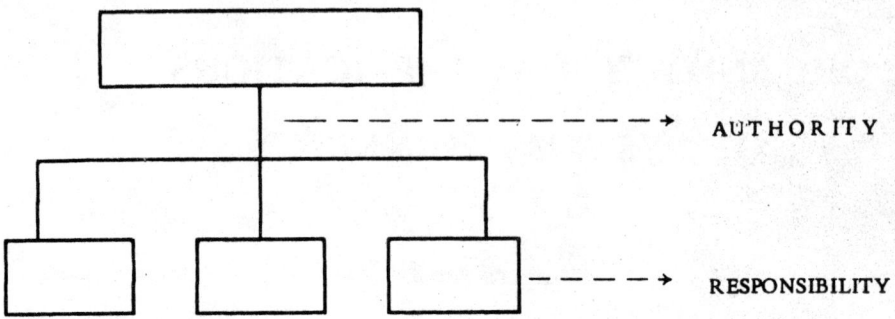

Figure 1.

The criteria used for dividing the whole into areas of responsibility, and determining their relative importance (line of authority) represent the major differences among organization theories. These criteria have evolved primarily around the five compoentns of a system.

- Input: functions and support activities required.
- Output: products and services produced.
- Environment: markets served.
- Process: transformation of input into output.
- Control: management and control - feedback.

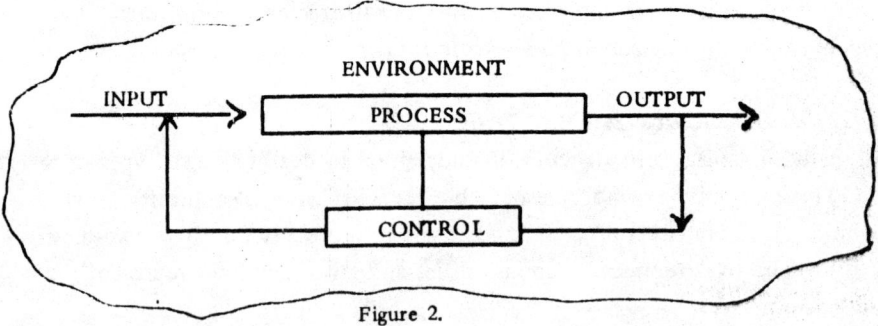

Figure 2.

The evaluation of organization theories has been a process of discovery of these five components, one at a time.

Unfortunately, recognition of each of the above components has resulted in a preoccupation with that component at the cost of others. It seems that a self-imposed unidimensional concept of organization, somehow, has prevented the development and realization of a multidimensional design.

ORGANIZATIONAL IMPLICATIONS OF SYSTEMS THINKING

Based on which dimension is emphasized, it is possible to distinguish five eras in organization development: production, marketing, research and development, humanization, and finally the systems era.

Production ERA	Marketing ERA	Research and Development ERA	Humanization ERA	Systems ERA
Input Orientation	Environment Orientation	Output Orientation	Process Orientation	Holistic and Interactive Approach
Efficiency	Growth		QWL	Development

The input orientation of the *production era* is manifested in its mechanistic view of the world and its sole concern with efficiency of production. Process is simply considered as the fixed, predetermined transformation of input into output, and output level invariably depends on inputs. The design criteria for such organization is dominated by the requirement for specialization (functional division of tasks), centralization (unity of command) and no deviation (tight supervision and control), representing the classical model of management. In this era organizations are treated as closed systems with no feedback from the environment. Bureaucracies are the prime example of this type of organization. The typical organization structure is functional and the only criterion for success is ability to produce.

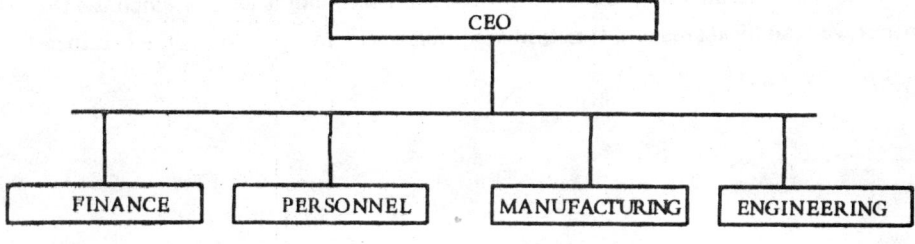

Figure 3.

The success of the production era in the mass production of goods and services and the resulting competition for markets led to a recognition of the influence of the environment on the organization. With the need to attract new customers, there has been a gradual shift of emphasis from manufacturing to marketing. The market-oriented organization is looked on as a living organism struggling for survival through growth. This pre-occupation with growth dominates the thinking of management. The division of labor also reflects this emphasis on the environment; the organizational structure's first priority is geography or market areas, function is secondary.

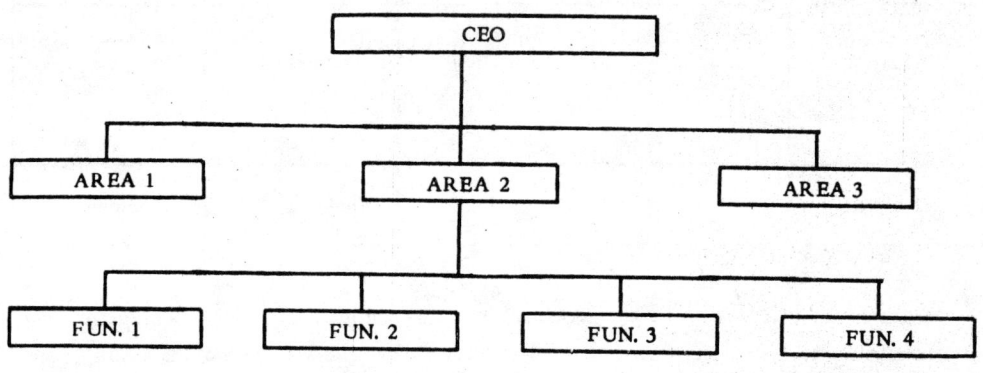

Figure 4.

The first level units (markets) are designated profit centers and the second level units (manufacturing), cost centers. This usually results in inefficient manufacturing units with output that is noncompetitive in cost and quality. Often this inefficiency is rooted in the failure to maintain state-of-the-art technology in order to ensure low costs.

The postwar scientific revolution, which led to organized research as well as the ever increasing pressure of competition, was responsible for the emergence of the *R&D era* with its output orientation. Product development, diversification, and a structure made up of divisions defined by products are a clear indication of a shift of emphasis toward output; geographical area and function are given a secondary status in the organization.

ORGANIZATIONAL IMPLICATIONS OF SYSTEMS THINKING

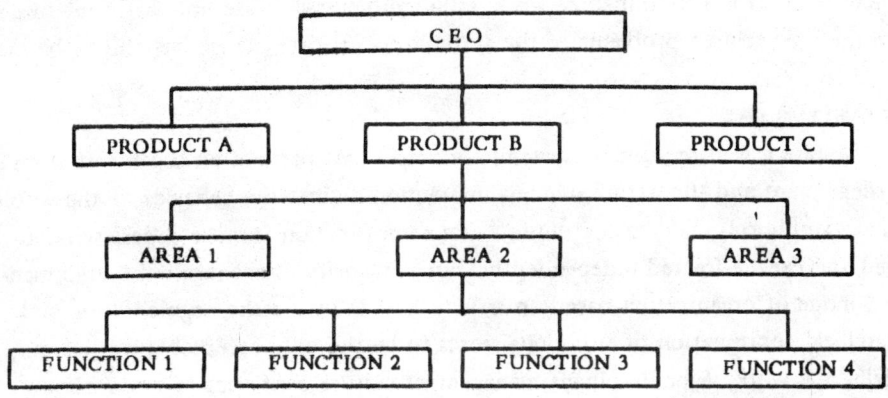

Figure 5.

This structure unfortunately, is not a proper response to the requirements of the R&D era. On the contrary it constitutes its core problem.

A proper approach to the challenges of the R/D era must recognize the following facts.

(1) Product life cycles are increasingly shortened; therefore, the future of the organization should not be solely dependent on the life cycle of a given product.

(2) It is not the product that makes the organization; it is the organization that makes the product.

Effective product development demands that product managers act as enterpreneurs and not be burdened by responsibilities for fixed production facilities; otherwise they will not be free to consider the development of a new product which may require different processes or facilities.

By locking the major divisions of an organization into a single product, the product-oriented structure ties the fate of the whole organization on that product. The organization, then, like the product, experiences periods of growth, maturity and inevitable decline.

Decline of the product causes insecurity, anxiety and unrest. The organization loses its most valuable resource, its people. Those who can leave do so, and those who remain become alienated.

Finally, the increasing rate of change, decreasing product life cycle, and subsequent increased insecurity and social unrest are the reasons for the emergence of the humanization era. The informal networks made up by the interaction among people in the organization is seen as determining the functioning of the firm. The formal structure of the organizations loses its importance and the process is emphasized. The recent management literature is full of concern for participation, democratization, quality of work life and the like. The efforts to humanize work, although necessary, are not sufficient to dissolve the inter-related problems of the systems era. This challenge has still to be met.

THE SYSTEMS ERA

Despite a well recognized systemic principle that parts of an organization are interdependent and the nature of their interaction defines the behavior of the whole, we have stubbornly held to the illusion that each function is more or less self-contained and can be treated independently. For a majority of managers the unidimensional mode of organization based on structurally defined tasks, segmentation and hierarchical coordination of functions seems to be the only acceptable way of organizing the work. A predominant management culture continues to value narrow specialization very dearly and considers any forms of redundancy in the organizational structure as wasteful and inefficient.

The traditional conception tends to create dichotomies which confront the management with a choice between opposing tendencies and orientations, for example: centralization or decentralization. In reality, however, different problem situations require different organizational capabilities. There are cases in which decentralized decision making provides the best answer and satisfies the requirement for responsiveness. But there are others in which some kind of centralization may be necessary to deal with the problem of coordination. Accelerating change and need for more flexibility forces large organizations into periodic and usually disruptive reorganization. On the other hand, the cost of reorganizations, the frustrations and tensions associated with them generates a desire for stability resulting in bureaucratic tendencies and resistance to change.

	Need for:	Leads to:
Short-term	responsiveness coordination	decentralization centralization
Long-term	flexibility stability	periodic reorganization bureaucratization

The survival of any organization in a turbulent environment depends on its ability to actively adapt to the changing needs of its members and its environment. The ability to adapt requires some forms of flexibility and responsiveness, which in turn demands that some degree of redundancy be built into the system.

This required level of flexibility and redundancy in an organization can be achieved by creating a multilevel, modular structure embedded in a multidimensional scheme. The organization, so conceived, then becomes capable of expanding or contracting by addition or deletion of replicable modules which have the means of vertical and horizontal interactions. The resulting mode of organization thus is capable of redesigning its structure and redefining its functions, so that it can exhibit different behaviors and produce different outcomes in the same or different environments. This means work can be organized in a variety of ways and indeed an organizational choice does exist.

The systemic answer to the humanization era, quality of work life and the problems of human effectiveness lies in the recognition that organizations are purposeful systems that are part of larger purposeful wholes, and they contain people who are also purposeful. Unless an organization effectively serves the purposes of its containing systems and its purposeful parts, they will not serve it well. This requires that the organizations be designed in such a way as to enable the parts to operate as independent systems with the ability to be relatively selfcontrolling while acting as responsible parts of a coherent whole that has the right to make collective choices. This can only be accomplished when fulfilling the needs of a higher system is an integral part of the goal of the lower system and vice versa. In this way each level achieves its preferred outcomes when the level above and the level below achieve theirs as well.[3]

Three factors that play an important part in the effective organization of multilevel, purposeful, information-bonded systems are identified by Boulding[4] as *Role*, *Exchange*, and *Threat*.

ROLE

As the unit of the social system role becomes an important factor in defining the behavior of a purposeful system with purposeful parts, the unit of these systems (family, group, organization, nation) is not so much the individual but the role imparted to him. Under different sets of circumstances and in different social settings individuals display different behavior. A good friend is not necessarily a good employee, a successful vice-president might make a poor president. This stems from their perceived role in that particular environment. The nature of these roles is influenced by expectations and limitations imposed by the social structure, the culture, and various environmental realities mapped by individuals.

Creation of proper roles involves a participative process in determining the ENDS as well as the MEANS. This process will enhance the creating of a shared image of the desired future and will result in a commitment to its pursuit and a sense of belonging. This collective commitment is required if the hierarchy of multilevel purposeful systems is to function properly. Therefore, participation is not a luxury but a necessity. Central to this notion of participation is one's ability to influence the system's behavior. There is no real participation if there is no sharing of power, and the key to sharing of power is the decentralization of control over the resources.

In this context, the underlying causes of rolelessness (*alienation*) can be explored in one or in a combination of the following factors, each corresponding to one of the five dimensions of the social system (power, knowledge, beauty, wealth, and values).[5]

Powerlessness. Powerlessness is equivalent to ineffectualness and impotence. When an individual feels that his contributions to the group's achievements are insignificant, or when he feels powerless to play an effective role in the system's performance, a feeling of indifference gradually sets in and he becomes alienated from the very system of which he is supposed to be a part.

Incompetence. Lack of sufficient knowledge and proper professional skills to carry out responsibilities of a specific role results in excessive anxiety and frustration.

Meaninglessness. Lack of a meaningful, exciting and challenging mission in life, suppression of an individual's need for creativity and achievement, and finally insensitivity toward the recreational aspect of the production process are among the main causes of meaninglessness.

Exploitation. A feeling of injustice in the fair distribution of a system's achievements is another factor which can cause alienation. When an individual feels that he has somehow been deprived of his fair share of recognition for contributing to a system's achievements, he becomes alienated and frustration will result.

Conflict. Finally, conflicting values within a social system contribute to alienation of its members. As mentioned before, the extent to which an individual's value image coincides with the "shared image" of his community, determines the degree of his membership.

EXCHANGE

Exchange represents all mechanisms that make the fulfillment of the needs of a system dependent on fulfillment of the need of the larger system of which it is a part, and vice versa.

Although it is possible to persuade purposeful members of a purposeful social system to engage in sacrifices for limited periods of time, it is highly improbable that they will accept this as a way of life.

Currently, under the assumption of a zero-sum-game, the exchange mechanism is based on a win/lose struggle. But in the reality of the emerging complex and highly differentiated social systems the win/lose struggle, despite its success so far, is being converted to a lose/lose one.

Nowadays, winning requires much greater ability than ever before. It has become

ORGANIZATIONAL IMPLICATIONS OF SYSTEMS THINKING

easier for any group to prevent others from winning than to win itself. Increasing numbers of small special interest groups are diluting the strength of the traditional power centers. Even many disadvantaged minorities have been forced to learn how to prevent the opposing sides from winning. But the illusion that increasing losses for the other side is equivalent to winning is the reason for prolonging the struggle and playing the game to a lose/lose end.

The central requirement for a systemic concept of exchange is the creation of a proper incentive system, a win/win environment within which the individual's struggle for his own gain will be enhanced by the degree of contribution he makes toward the satisfaction of the needs of the higher system and those of his fellow members.

For example: consider an organization in which all units are profit centers. Each unit retains a percentage (suppose 5%) of its profit after meeting certain requirements as an incentive to be distributed among its members. This percentage would increase to 7.5% if all of the peer units achieve their goals; and would double to 10% if the larger system of which it is a part achieves its desired outcome as well. This system was instituted in a government agency just converted to a profit center. The firm, engaged in management consultancy, research and education, grew from a three million to a sixty million operation in less than five years and enjoyed a six-month queue for its services.[6]

THREAT

Threat covers criteria which make the membership in a system dependent on avoiding a certain set of behaviors considered antagonistic to the survival of the larger system of which it is part, and vice versa.

A participant in a decision must be willing to live with the consequences of the decision. Those who have no stake in the outcome or the viability of a system cannot be entrusted with the final decision about that system. This means that authority and responsibility must go hand in hand. One of the pathologies of bureaucratic systems is produced by the fact that a tenured bureaucrat can make decisions which affect others with no direct consequence to himself.

Note that each of these factors — role, exchange, threat — influences certain behaviors in its own way. Where role is of central significance exchange might not be effective; where exchange can motivate a certain behavior, creating a sense of threat might exhaust opportunities that could be used in more critical situations. Therefore, they are not substitutes but complements.

The rest of this chapter is devoted to a detailed description of an interactive model of social organization. The model not only is capable of producing redundancy along with efficiency, but also effectively combines flexibility with stability, responsiveness with coordination and finally participation with responsibility. The model focuses explicitly on interactions between different parts of the system. It differentiates and explains the relationship between various types of functional units, and clarifies the roles that ought to be played by each purposeful unit within the organization. The participative style of decision making proposed in the model is also compatible with the increasing demand for humanization of work and the emerging concept of quality of work life.

2. MULTIDIMENSIONAL MODULAR DESIGN

The following design is based on the interactive approach of the systems era and the recognition that the three common criteria — input (functions), output (products) and environment (markets) — traditionally used to set the priorities for division of the labor, and establishment of the line of command, are complements. Treating them as independent dimensions and managing their interactions results in a system that is more than the sum of its parts. The multidimensional design eliminates the need for periodic reorganization when a change in environmental conditions requires a change of emphasis from one orientation to another; for example from market to product orientation. The assumption that a change in strategy of an organization would invariably require a change in its structure is based on the traditional and unidimensional concept of structure.

The assumption that a given structure can only be effective in production of a given function has been successfully challenged by Russell Ackoff. In his work on purposeful systems, Ackoff shows that different structures can produce the same outcome and that the same or similar structures are capable of producing different outcomes (Figure 6).

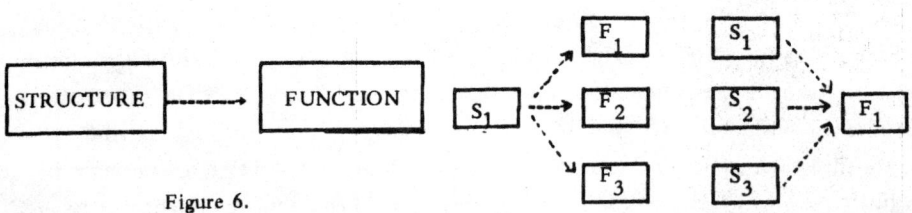

Figure 6.

To achieve this flexibility, each dimension of an organization in the multidimensional design corresponds to one of the components of a system as shown in Figure 7.

ORGANIZATIONAL IMPLICATIONS OF SYSTEMS THINKING

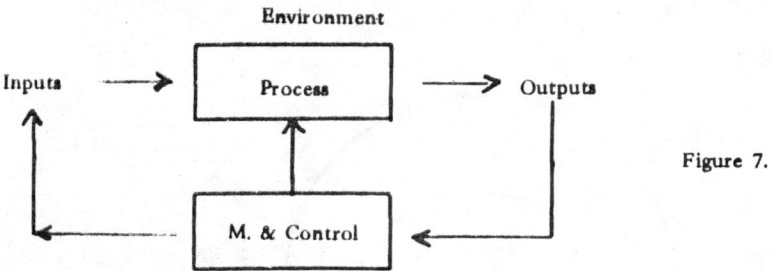

Figure 7.

Outputs are the goods or services produced; inputs are the resources and facilities required. The environment defines the characteristics of markets and communities served. Process represents the mode of organization, style of decision making and the rules of transformation; while management and control provides for horizontal coordination and vertical integration of all activities and the learning system for effective development of the organization.

The functions and specifications for each of the dimensions are as follows:

OUTPUT. Responsibility for achieving an organization's end is vested in the output dimension. This is where the organization actually happens. The output dimension consists of a series of *general purpose*, semi-autonomous and ideally self-sufficient units charged with all of the activities ultimately responsible for achievement of an organization's mission and production of its outputs.

Note that the semi-autonomous, self-sufficient and purposeful units, for simplicity, are referred to as modules. Modules are self-sufficient and autonomous to the degree that the integrity of the whole system is not compromised.

Output modules have the following characteristics:
- Each output module represents a specific level in the hierarchy of multilevel purposeful systems.
- Each output module is a miniature of the whole — the larger system of which it is a part — and may contain lower-level modules (sub-modules) as its own parts.
- Each output module is considered to be the environment of its corresponding sub-modules. Since each module consumes scarce resources of its environment, its outputs should be responsive to the needs of the environment.

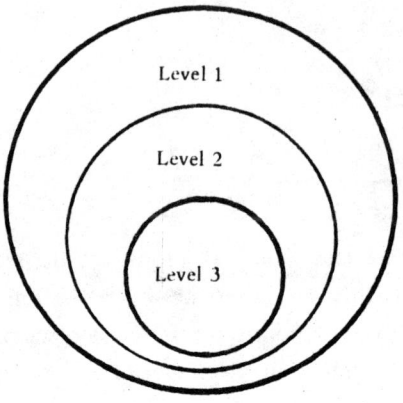

Figure 8.

- The lowest-level output module is the smallest unit that can be accountable for the production of a tangible and measureable output.
- Performance of an output module is, preferably, as insensitive to the behavior of other peer output modules as possible; it has enough authority over its resources – money, people, facilities – to be responsible and accountable for its success or failure.
- Each output module will have an organizational structure (multidimensional) very similar to the larger system of which it is a part.
- Each module is allowed to retain a percentage of its contributions above a minimum level to be used for incentive and internal development.
- Each output module is responsible for making those decisions which affect only its operations. Decisions that impact on the other units will be made at higher levels with participation of all affected modules.
- An output module is usually conceived as a unit hosting a product, a project, or a program. A product module has an entrepreneurial role. It has the managerial responsibility for the development, design, production, marketing and finally the distribution of a given product.

It has the ultimate responsibility for technological feasibility, marketability and profitability of the final product.

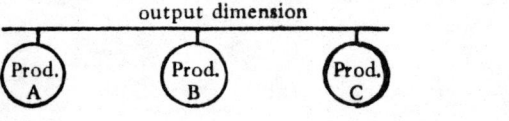

Figure 9.

General purpose output modules

ORGANIZATIONAL IMPLICATIONS OF SYSTEMS THINKING

INPUT DIMENSION. To take advantage of the synergy effect specifically with regard to economies of the scale, and the need for specialization, some of the functions and support services required by output models can be shared. These shared services and specialized functions can be provided by groups of special purpose modules which together constitute the input dimension of the organization.

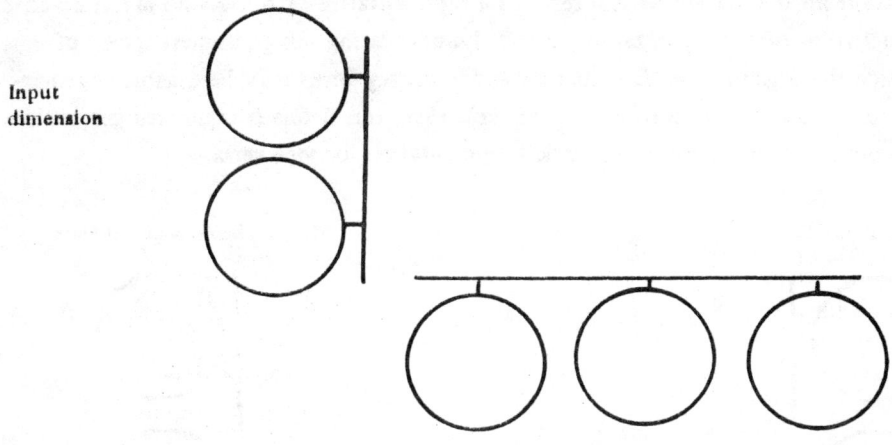

Figure 10.

For example, in terms of the product-management concept discussed above, the manufacturing, engineering, maintenance, information and data processing, and personnel functions might be shared by output modules.

Designating the manufacturing unit as a profit center in the input dimension not only results in more competitive and flexible facility management, but also provides the product managers with the freedom to buy their manufacturing requirements from inside or outside of the organization without being constrained by fixed facilities.

The input modules, in general, are provided with a working capital and are expected to earn their operating expenses plus a return on the investment by charging the market price for their services. If insufficiency or unpredictability of demand makes it necessary to provide additional support for a function, then the rule, in general, is to subsidize the demand instead of subsidizing the supply. In the early stages of the conversion, total operating budget of an input unit may be given to it by means of a purchase contract for its total services.

ENVIRONMENTAL UNITS. Every organization utilizes scarce resources of the environment, therefore, its outputs ought to be responsive to the needs of that environment. Interaction with the environment is facilitated by the environmental modules, the third dimension of the organization. Two main functions of these modules are: (1) distribution and (2) advocacy. Distribution units perform the sales function, that is, attracting the clients, and representing the organization to outside.

Advocacy is responsible for sensing environmental conditions and exploring expectations from the system. It will serve as a representative of those who are affected by the activities of the organization, especially, advocating the consumers' point of view inside the organization. Distribution and advocacy units may be organized geographically or by segmentation of a given market. However, if one is organized geographically the other should be based on market segmentation, or vice versa.

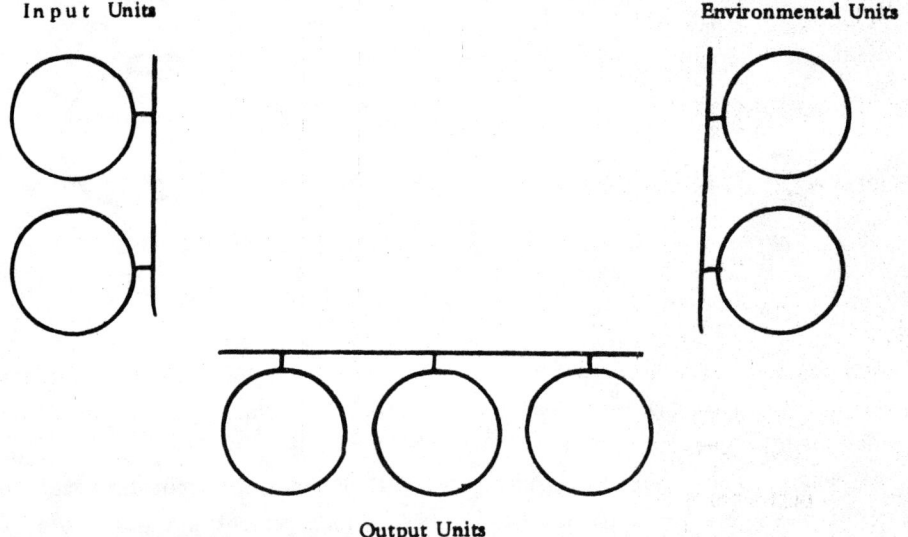

Relationships between input, output and environmental units

Defining the relationships between input, output and environmental modules is the most critical task of this conception. Extreme difficulties are encountered when several output units share the vital services of a bureaucratic core that is an input unit set up as an overhead center, then the relationships become unclear and ambiguous. The problem is aggravated when performance criterions are also implicit and conflicting.

The key problem of matrix organizations also has been their failure to meet the challenge of defining and managing the implicit relationships between the network of input and output units. The "two-boss system" not only has failed to dissolve the problem, but has resulted in further confusion and frustration.

In the proposed model the answer for this inherent complexity is to create an in-

ORGANIZATIONAL IMPLICATIONS OF SYSTEMS THINKING

ternal market environment, so that the relationships between input, output and environmental units are converted to the same type of relationships that exist between a supplier, a producer and a distributor. To given an organization a market orientation requires that each part of an organization consider the marketing consequences of what it does. Such an orientation is best obtained by creating a market within the organization. Creation of an internal market not only eliminates growing problems of bureaucratization, but also provides an effective means for dealing with allocation and evaluation problems. It also makes it possible to evaluate the performance of every unit at every level in exactly the same way. In the multidimensional design, this can easily be achieved by gradual replacement of cost centers by profit centers.

This means that input, output and environmental modules become profit or performance centers. Of course, this alone cannot dissolve the problem because, if there is only one input unit (for example a computer center) which provides the required services for all of the output modules within the organization, then the input unit has a monopolistic advantage, while output units are in a no-choice situation. To avoid this undesirable situation, the modules ought to have a choice with respect to selling or buying their required services from inside or outside the organization.

PROCESS. The decision making process of an organization is reflected in its mode of planning. In the proposed model this is done by planning boards, the fourth dimension of the organization.

Planning as traditionally practiced is one or a combination of two dominant types: *reactive* and *preactive*.[7]

Reactive Planning is concerned with *identification* of *deficiencies* and design of projects and strategies to *remove* or suppress them. It deals with parts of an organization independently of each other.

An organization is, however, a system whose major deficiencies arise from the way its parts interact, not from the actions of its parts taken separately. Therefore, it is possible, and even likely, that improvement of the performance of each part of an organization taken separately will result in a deterioration of the performance of the organization as a whole.

Preactive planning consists of two major activities: *prediction* and *preparation*. The objective is to forecast the future and then prepare the organization for it as well as possible.

Unfortunately such forecasts are chronically in error since the social, economic and political conditions, as well as the behavior of supplier, consumer and competitor behavior are affected by what the planned-for organization, and others like it, do. Therefore, it is precisely such plans taken together that shape the future.

Systems methodology rests on the *interactive* type of planning, which is based on the assumption that the future is created by what we and others do between now and then. Therefore, the objective is the *design* of a desirable future (idealization); and the *invention* or selection of ways of bringing it about (realization).

Interactive planning is a multi-level participative process. In each level a team of all major stakeholders of the system is formed. They start with the assumption that their system was destroyed last night but everything else in the environment, including the containing system, is still in existence. The task before them, then, is to ideally redesign the whole sub-system from scratch.

An ideally redesigned system is one with which the designers would *now* replace the existing system if they were free to replace it with any system they wanted.

Constraints. Such redesign is subject to only two constraints:

1. The design must be *technologically feasible*; that is, it cannot incorporate any technology that is not known to be feasible at the time the design is produced.
2. The design must be *operationally viable*; that is, it must be capable of operating in the current environment of the system planned for. However, no consideration should be given to the feasibility of implementing the ideally redesigned system because doing this constrains creativity. Moreover, the total design may be feasible even though it contains parts that are infeasible when considered separately.

The idealized design process has three steps: selecting a mission, specifying desired properties of the system, and designing the ideal-seeking-system.

Mission

A mission is an overriding purpose that can unify and mobilize all parts of the organization planned for. The formulation of the mission can be an exciting challenge to virtually every stakeholder. It can also provide a focus for the design process.

The mission statement basically identifies the reason for systems existence and deals with the "WHY" question. The statement should also specify what effects systems wants to have on each class of its stakeholders.

An organization is a purposeful system with purposeful parts and is itself contained in a larger purposeful system. Its main function is to serve both the purposes of its members as well as those of its environment. Therefore, an organization's mission should address (a) itself as a purposeful system, (b) its role in relation to those within the organization, and (c) the role it ought to play in the environment.

ORGANIZATIONAL IMPLICATIONS OF SYSTEMS THINKING

Specifications

Specifications are the desired charactdristic of the system. They basically deal with the "WHAT" questions. They should address the following four categories:

1. Environment: What is the need in the environment that systems is trying to fulfill? Who are the main actors and what variable they control? What are their expectations?
2. Outputs: What product or services the system should offer? and, What should be their special characteristics?
3. Inputs: What resources such as material, facilities, money, information, people system needs in order to provide the specified outputs, and what should be their characteristics?
4. Process: What kind of decision making process is desired? What should the performance criteria of the system be? and What style of governance should it have?

Design

The design deals with the "HOW" question: How is the system to ensure that mission will be achieved and the specifications realized? In order to answer this question an appropriate structure and process must be designed.

Structure: How the system is going to be organized? What will be its components and their relationship?

Process: How will policies and procedures be developed, organized, and carried out? How will production of goods and services be planned and scheduled? and finally, How will learning and adaptation, coordination and integration be realized? and, How will the systems performance be evaluated?

PRINCIPLES. Three principles should be followed in the idealized-design process:

1. Where there is no objective basis for making a design decision, the system should be designed so that it can determine experimentally which of the available alternatives is the best.
(This applies to properties on which consensus is not reached.)
For example, if the designers have no basis for deciding which of two possible new practices to include in their design, they should incorporate an experimental comparison of both.
2. The system should be designed so it can continuously evaluate features that have been designed into it and decisions that are made within it. This enables it to *learn* efficiently.

3. Since any design incorporates assumptions about the future, the system should be designed to monitor these assumptions and to modify itself appropriately when an assumption turns out to be false. This enables it to *adapt* effectively.

Therefore, the product of an idealized design is an *adaptive-learning system*. The output is neither utopian nor ideal because it is subject to improvement. It is the best *ideal-seeking system* that its designers can conceptualize *now*, but not necessarily later.

Details of interactive planning can be found in the "Guide to Interactive Planning."[8] The rest of this chapter, however, deals only with the organizational aspect of the planning process.

The planning boards are the main organizational vehicle for the planning function. They are created at every level, within all units, to provide for integration and coordination of input, output and environmental units. In general, each board consists of (1) the manager of the unit whose board it is, (2) his immediate superior, and (3) his immediate subordinates. Note that subordinate units are categorized as input, output and environmental units.

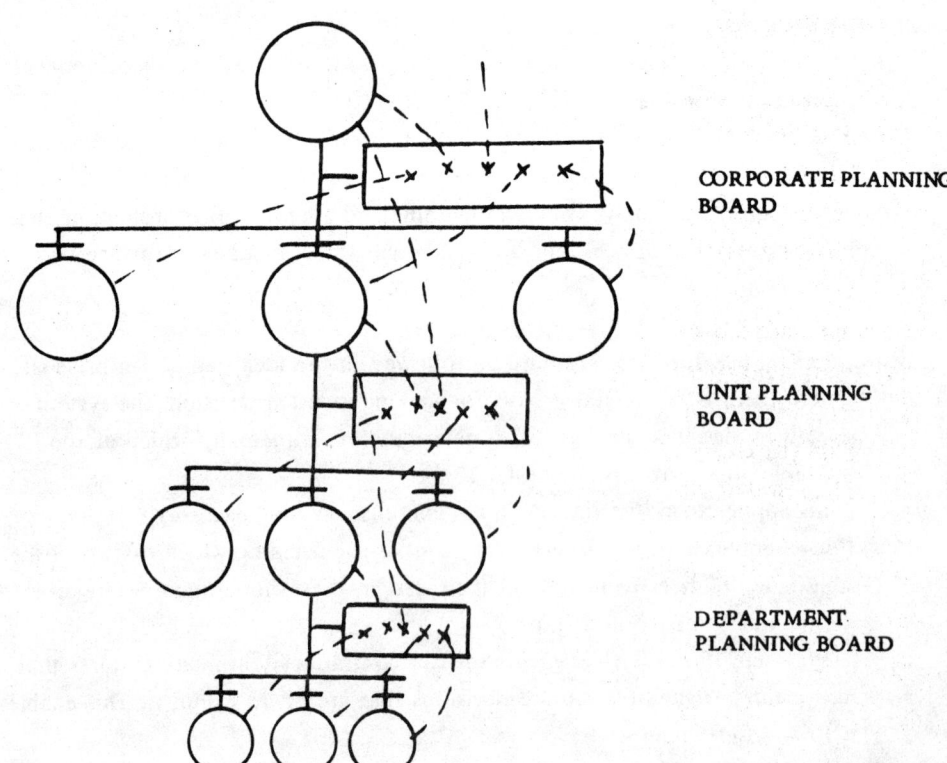

CORPORATE PLANNING BOARD

UNIT PLANNING BOARD

DEPARTMENT PLANNING BOARD

ORGANIZATIONAL IMPLICATIONS OF SYSTEMS THINKING

Presence of all managers of peer modules in the planning board of their respective boss provides a mechanism for horizontal coordination of their activities. Participation of a manager in superior and subordinate boards and his consequent interaction with two levels above and two levels below results in a vertical integration of organization and a responsive and flexible decision process.

Each planning board is so designed as to allow maximum autonomy for decision making without compromising the integrity of whole system and the positive synergy that might exist among its components. The planning boards are policymaking bodies, not advisory committees. Their repsonsibility is to formulate the decision criteria and redesign their system as needed. If consensus cannot be reached on the formulation of a policy or development of a strategy, then an experiment is designed to dissolve the conflict. All policy decisions that affect only one particular module are made by the planning board of that module, but any decision that affects more than one module will be made at the next higher level.

Every module which has a planning board must have a mission and a clear, operationalized measure of performance. This need not be formulated solely in monetary terms. Once the decision criteria are formulated the executive decisions are made by respective managers.

In formulating decision criteria, underlying assumptions and expected outcomes of every policy decision must be explicitly specified. Constant monitoring of these assumptions and the comparison of the actual outcomes versus expected ones is the basic requirement for creating a learning and adaptive system. The system that relies more on experimentation than experience to deal with present and future uncertainties, has the ability and willingness to redefine problems and solutions based on emerging realities.

MANAGEMENT AND CONTROL

The fifth dimension of an organization is its management and control system. It represents the executive function of the organization which oversees the operation of the whole system and orchestrates the activities of all other dimensions.

The executive function has the responsibility to create a vision, generate a shared image of a desired future and provide the leadership for achievement of organizational mission. It has the final responsibility for financial viability, technological ability, and human effectiveness of the organization as a whole. A small group of specialized staff divided into three areas of financial systems, human systems, and technical systems will assist in discharging these responsibilities.

The *financial systems* group is engaged in those systematic efforts directed at con-controlling the resources and monitoring the financial performance of the organization.

The *human systems* group is involved in activities directed at development of human resources and monitoring the quality of work life within the organization.

The *technical systems* group is responsible for those activities directed at achieving a technological advantage, monitoring the state of the art, and controlling the quality of the output of all operating systems.

Technical, financial, and human systems groups provide the staff support for the planning board.

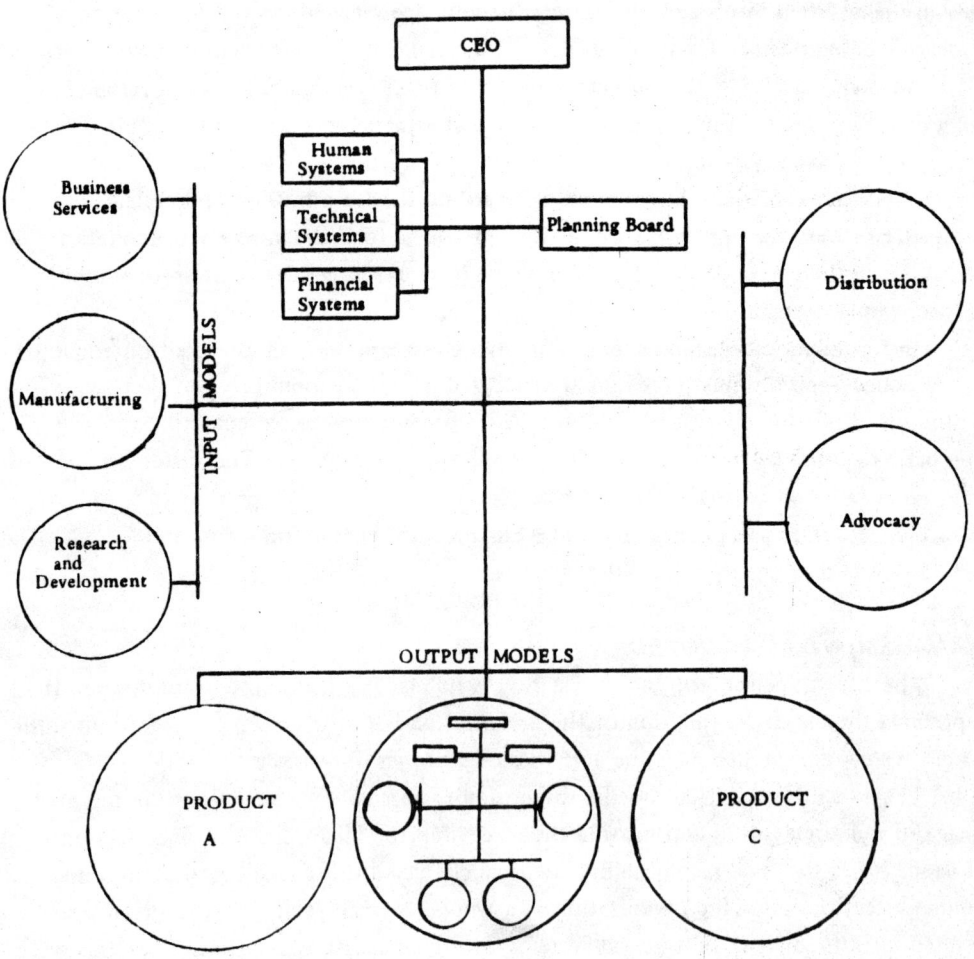

NOTES

1. See J. Gharajedaghi, "On the Nature of Development," Human Systems Management, 4 (1984).

2. The following is a partial list of the various types of organizations that have used this concept:
    Manufacturing sector: heavy equipment manufacturer, machine tool manufacturer, home appliances, steel mills, copper and aluminum industries.
    Service organizations: banking, retail, contracting, insurance.
    Professional organizations: research, consultancy, education, engineering, hospital and medical centers.
    Government: ministries of education, agriculture and health.

3. R. Ackoff and F. Emery, *On Purposeful Systems*, Intersystems Publications, Seaside, 1981.

4. K. Boulding, *Beyond Economics*, University of Michigan Press, Ann Arbor, 1970, pp. 43-53.

5. See J. Gharajedaghi, "Obstructions to Development," Human Systems Management, 4 (1984).

6. For further elaboration of this concept of exchange and its application on a national level, see J. Gharajedaghi, "Toward a New Social Calculus,"
Social Systems Sciences Department, University of Pennsylvania, 1982.

7. R. Ackoff, *Creating the Corporate Future*, John Wiley, New York, 1981.

8. R. Ackoff, E. Vergara, J. Gharajedaghi, *Guide to Controlling Your Corporate Future*, John Wiley, New York, 1984.

7
PERFORMANCE CRITERIA AS A MEANS OF SOCIAL INTEGRATION*

with ALI GERANMAYEH

In contrast to machines and organisms in which integration of the parts into a cohesive whole is a one time proposition, for social systems, the problem of integration is a constant struggle and a continuous process. In seeking effective social integration, compatibility must be continuously recreated, first, between the different levels of purposeful systems (vertical), and second, among purposeful members at the same level (horizontal). A third kind of compatibility should be included, a continuing concern for the interests of past and future stakeholders in a social system (diachronic). In this context, the dichotomy of individuality and collectivity poses a dilemma that cannot be resolved under current sets of assumptions and traditional concepts of organization.

As discussed extensively elsewhere,[5] there are three different ways of looking at organizations: mechanistic, organismic, or as a social system. In the mechanistic view organizations are considered to be instruments of their owners and efficiency, reliability, and predictability to be their main measures of performance. A machine with passive, nondeviating parts is a model for such systems.

The organismic view conceptualizes organizations as living systems with the parts as organs having essential functions but no purposes of their own. Environment is also seen as purposeless and passive provider of necessary inputs. Thus, exploitation of the parts and of the environment for the maintenance and growth of an unstable structure becomes the main preoccupation of the system.

The social systems view considers organizations as voluntary associations of purposeful members — systems that manifest a choice of both ends and means. Therefore, the purpose of an organization is to serve the purposes of its members while at the same time serving the purposes of its environment.

The presence of an element of choice in the behavior of the members places the

---
*This paper was written as a part of a festschrift in honor of Russel L. Ackoff and is reproduced here by permission of the editors.

social system in a class by itself. Lack of appreciation for the implications of the factor of choice in the behavior of members of social systems becomes the main source of the confusion and dilemmas encountered in organizations conceived as mechanistic or organismic systems.

In the context of purposeful behavior, the relationship of an individual with the larger system of which he is a member is critical. Despite a desire for individuality and uniqueness, individuals also display a strong tendency toward conformity and a desire to identify with, and to be members of, groups of their choice.

To the extent that membership in a group is desired, the performance criteria by which the group evaluates the behavior of its members have a profound effect on the manner in which individuals conduct themselves. The degree of the influence, however, depends on the extent to which the performance criteria: (1) represent collective choice, (2) are legitimized by the culture, and (3) are compatible with the performance criteria of the other groups of which the individual is a willing member.

In this chapter we will deal with the concept of social integration by reflecting on the roles of performance criteria, performance measures, and incentive systems, and we will propose a framework to synthesize these concerns in the design of organizations.

Vertical Compatibility

As an open purposeful system an organization is part of a larger purposeful system, the nation. At the same time it has purposeful individuals as its own members. These create a hierarchy of purposeful systems of three distinct levels. These three levels are so interconnected that an optimal solution can not be found at one level independent of the other two.[1] Therefore, effective integration of multilevel purposeful systems cannot be achieved without performance criteria that make the fulfillment of the needs and desires of a purposeful system dependent on fulfillment of the needs of the larger system of which it is a part and vice versa.

This means that the implicit sets of zero-sum assumptions ought to be challenged and a win/win environment created so that the efforts of individuals for their own gain are enhanced by the degree of contribution they make toward satisfying the needs of the higher system of which they are willing members. This would change the measure of success and relative advantage of various activities for the actors in favor of those activities that collectively optimize the tri-level objectives of purposeful systems.

If organizations are to serve their members as well as their containing whole, they must be able to dissolve the conflicts among them. In contrast to solving or resolving, which is used whenever the conflict situation is conceived to be unidimensional (a zero-sum game) to dissolve a conflict is to change the nature and/or the environment of the entity in which it is embedded so as to remove the conflict. This is a multidimensional

PERFORMANCE CRITERIA AS A MEANS OF SOCIAL INTEGRATION

conception of conflict, characterizing a non-zero-sum situation.

This concept is based on the recognition that sets of opposing tendencies, usually treated as dichotomies, are in fact complementaries. They coexist and interact continuously so that the relationship between opposing pairs might be characterized by an "AND" rather than an "OR" relationship.

For example, one of the prime functions of social systems, the production and distribution of wealth, is a constant source of conflict which results in the dichotomies of collectivity and individuality, market economy and planned economy, and so on. However, production and distribution of wealth form a complementary pair. Without an effective production system there can never be an effective distribution system. To fail to note this important interdependency is to leave out the most important challenge of the problem. An obsession with distribution without a proper concern about production, will result in nothing but equitable distribution of poverty. This is not to undervalue the importance of an equitable distribution system, but to emphasize its inseparable connection with production.

The answer is not a compromise, but a synthesis. This synthesis can be approached only when production and distribution of wealth happens at the same time.

Consider the following simple exchange system:

A productive unit consumes the scarce resources of its environment; in return it produces outputs (goods or services) that partially fulfill the needs of that environment. The assumption is that the unit will survive as long as the total value of the outputs produced is greater than or equal to the total value of the inputs it consumes. The pricing system determined by "dollar votes" is supposed to be a reliable and sufficient criterion for determining production and distribution priorities. This supposition might be tenable if (1) dollar votes were more equitably distributed, and (2) end-prices were not manipulated. However, factors such as price control or government protection make the actual cost of service much higher than perceived. In other words, inputs are purchased from the environment at a lower price and outputs (measured by the classical accounting method) are made to look more valuable than they really are. Furthermore, despite the fact that creation of employment opportunity for all members of a social system is an effective means of simultaneous production and distribution of wealth, existing social calculus considers employment only as a cost, therefore, and not surprisingly, tries to minimize it.

To remedy the situation we need a new framework, one that will use employment on both sides of the equation, input as well as output. We also need a performance criteria that in addition to efficient production of wealth explicitly considers its proper distribution as a social service to be adequately rewarded.

The following scheme is a simplified version of an attempt to measure the actual costs and benefits of each major economic activity as perceived on the national level. It complements the productive strength of market economy by enhancing its allocation function. The model registers the needs of those members who lack the dollar vote to register their needs. It also explicitly values the distribution of wealth (salaries paid) as social service.

For simplicity let us limit inputs to the two categories of (1) raw materials, and (2) human resources, and the outputs to the two corresponding categories of (1) finished goods produced, and (2) employment opportunity created (this assumes that distribution of wealth is a social service). Assigning a "scarcity coefficient" to each set of inputs obtained from the environment, and "need coefficient" to each set of outputs (goods/services) yielded to the environment, we can compute the relative contribution of each major economic activity using the following table:

INPUT CONSUMED	OUTPUT PRODUCED
RAW MATERIAL	**GOODS PRODUCED**
A: Quantity(a) x Price(a) x Scarcity Coefficient(a) = VC(a)	D: Quantity(d) x Price(d) x Need Coefficient(d) = VP(d)
B: Quantity(b) x Price(b) x Scarcity Coefficient(b) = VC(b)	E: Quantity(e) x Price(e) x Need Coefficient(e) = VP(e)
C: Quantity(c) x Price(c) x Scarcity Coefficient(c) = VC(c)	F: Quantity(f) x Price(f) x Need Coefficient(f) = VP(f)
HUMAN RESOURCES UTILIZED	**EMPLOYMENT OPPORTUNITY CREATED**
Number of Employees (I) x Train. Cost/Productive yrs. x (Scarcity Coeff.) = VC(I)	No. of employees (I) x Salary paid x NCE = Vp(I)
Number of Employees (II) x Train. Cost/Productive yrs. x (Scarcity Coeff.) = VC(II)	No. of employees (II) x Salary paid x NCE = Vp(II)
Number of Employees (III) x Train. Cost/Productive yrs. x (Scarcity Coeff.) = VC(III)	No. of employees (III) x Salary paid x NCE = Vp(III)
Total value consumed: TVC	Total value produced: TVP
TOTAL VALUE PRODUCED/TOTAL VALUE CONSUMED = Contribution Ratio	

Category (I)   = Highly specialized
Category (II) = Skilled
Category (III) = Unskilled
Vc: Value Consumed
Vp: Value Produced
Train. Cost: average training cost of an individual in the category
Productive years: average number of productive years in the category
NCE: need coefficient for creation of employment in the category

PERFORMANCE CRITERIA AS A MEANS OF SOCIAL INTEGRATION

This conception of social calculus, if it has embedded in it an explicit incentive system to reward activities with higher social contributions will result in a more effective vertical integration.

Suppose a certain productive unit produces bread with a contribution ratio of 2, but the low rate of return on investment of 8% (because of the weak purchasing power of the consuming class). On the other hand, suppose another unit produces yoyos with a contribution ratio of 1, but the rate of return on investment of 18%. Then our incentive system ought to be able to change the relative rates of return on investment in favor of bread. An integrated and coordinated application of well known tools such as (1) differentiated loan structure, (2) differentiated interest rate structure, and (3) differentiated tax structure, will overcome the problem. Depending on the contribution ratios (computed from the previous table), a different loan/equity ratio, a different interest rate, and a different tax rate can be assigned to each major economic activity. This, as demonstrated by the following table, will increase the rate of return on investment for bread to 18% and decrease that of yoyos to 10%.

The advantage of such a scheme is that it will minimize the bureaucratic dangers associated with centralized planning while enhancing the strength of market economy by promoting a more equitable allocation and distribution system.

The method of determining scarcity and need coefficients is based on successive approximation utilizing interactive planning.[2] The details of this process, although a critical aspect of this conception, is beyond the scope of the present paper. However, it is important to note that the initial raw coefficients, are revised and updated regularly and continuously in light of further experiences.

PRODUCT	BREAD	CANDY	YOYO
Contribution Ratio	2	1.5	1
Current return on investment	8%	12%	18%
Equity	$1,000,000	$1,000,000	$1,000,000
Equity/loan ratio	1/4	1/2	1/1
Total loan	$4,000,000	$2,000,000	$1,000,000
Interest rate	5%	9%	16%
Cost of loan	$200,000	$180,000	$160,000
Total capital employed	$5,000,000	$3,000,000	$2,000,000
Income	$400,000	$360,000	$360,000
(Income-cost of loan)	$200,000	$180,000	$200,000
Tax coefficient	10%	25%	50%
Net Income after taxes	$180,000	$135,000	$100,000
Final return on equity	18%	13.5%	10%

CHANGING RELATIVE RATE OF RETURN ON INVESTMENT BASED ON CONTRIBUTION RATIO

Horizontal Compatibility

The development of major organizational theories is associated with increasing complexity and an emerging need for further differentiation. In response to requirements for specialization, the prime concern of every organizational theory is to define the criteria by which the whole is to be divided into functional parts. Most major organizational theories have implicitly assumed that the whole is nothing but the sum of its parts and have conveniently ignored the fact that effective differentiation requires incorporation of a mechanism that would integrate the differentiated parts into a cohesive whole. In this regard, the classical school of management, depends solely on the *unity of command* and the imperative of *no deviation*. At the opposite end, the neoclassical school, advocating decentralized structure, relies on the assumption that perfectly rational micro-decisions would automatically produce perfectly rational macro-conditions. More significantly, the assumption that it is necessary to differentiate among the criteria by which functional units are to be evaluated has led to the whole range of incompatible performance criteria embedded in the so called cost center, revenue center, overhead center, and so on.

Consider for example, a typical setup within a corporation. The performance criterion for manufacturing units is the minimization of cost of production. That of the marketing units is to maximize sales. (These units are often referred to as *cost centers* and *revenue centers*.) Intuitively one would expect that the interaction between the two centers be complementary and result in maximum efficiency. Unfortunately, this is not so. In fact, in most organizations the relationship between marketing and production is fraught with constant friction.

The reason is that this design violates a basic systemic principle. Optimization of each of the parts of a system in isolation will not lead to optimization of the system as a whole. The two objectives of cost minimization and revenue maximization, taken independently, lead to a basic contradiction within the system. In order to maximize sales, marketing will have to respond to market demands for a variety of products, customized features, change of delivery schedules on short notice, and so on.

Minimization of cost of production, on the other hand, is achieved through standardization of the production process, reduction in the number of products, regularization of the production schedules, and similar measures. Thus, the basic contradiction emerges: marketing can comply with its performance criteria only at the cost of manufacturing and vice versa.

Ironically, the only reason that this setup works in present-day organizations to the extent that it does, is that the performance criteria are not taken seriously. (This is the major advantage of systems with purposeful parts over mechanistic and organismic systems. Such incompatibility could never be tolerated in a mechanistic system.)

PERFORMANCE CRITERIA AS A MEANS OF SOCIAL INTEGRATION

The usual solution that most corporations adopt for this problem is one of compromise. The higher-level authority over both centers determines which set of criteria should dominate the other at any particular time. The dominance, of course, would oscillate between manufacturing and marketing in order to keep a sense of balance.

A totally different approach to this problem is to attempt to create compatibility of performance criteria rather than seek a compromise between incompatible sets. One way to achieve this is to change the performance criteria for marketing and for manufacturing so that they would both try to maximize the *difference* between cost and revenue. This concept can be operationalized in a profit center design where the relationship between marketing and manufacturing is based on *exchange*, much like that of a customer and a supplier. Both units are now expected to be value-adding operations.

Consider the difference in how the question of flexibility of delivery schedules is handled by the two designs. Flexibility is a value to some users for which they are willing to pay a certain premium. For a cost center, this premium has no value whatsoever. The only thing that is important to the cost center is that a change in production schedules will increase the cost. Since it is not concerned with revenues, the cost center will resist flexibility even when it results in positive net value to the corporation as a whole. Furthermore, since transfer of costs is based on average cost rather than marginal cost, there can be no distinction between users who demand flexibility (or various degrees of it) and those who do not; average cost is the same for both.

A profit center, on the other hand, examines each opportunity on a marginal cost versus marginal revenue basis. The price that the customer is willing to pay for the additional service is balanced against the marginal cost that the supplier will have to incur.

While a cost center is instinctively resistant to any change of operations requested by marketing, a profit center looks forward to opportunities to increase its net contribution to the system.

Note that in a profit center design it may still be possible for one unit to benefit at the cost of another. But the critical difference is that a win/win situation is now a possibility. Both production and marketing can benefit from meeting customer demands and more significantly the performance criteria for the units are compatible. Note that uniformity of performance criteria although desirable is not necessary, the important point is compatibility. That is to say that performance criteria must be designed in such a way that the success of one part does not imply failure of another.

Diachronic Compatibility

Concern for diachronic compatibility in a social system is concern for its continuity. Among stakeholders of an organization there are those who had been members of the system in the past and those who will be its members in the future. The argument for compatibility between the interest of past, present, and future members, espeically on ethical grounds, is so rich that it is beyond the scope of present paper; our concern here is essentially pragmatic.

It is not difficult to appreciate that a social system can succeed today, at the expense of its future, or suffer today in the creation of a better future. It can also be demonstrated that past members of a social system can have a profound (negative or positive) influence in shaping its present. However, although a need for compatibility between the interest of present and future members is more or less appreciated, the same need for compatibility between the interest of present and past members is not readily recognized.

In some cultures, interests of past members continue to dominate the present, while in other cultures there is no concern for the interests of those who are gone — out of sight, out of mind. Nevertheless, rejection of the interests of past members is as undesirable as acceptance of their dominance. The effectiveness of an organization, as a voluntary association of purposeful members, depends on the degree of their commitment and sense of belonging. In this context, alienation is a serious obstruction to an organization's development. Incompatibility between the interests of past, present, and future members is one of the main sources of alienation of its present members. This is so because a constant threat to the long-term viability of the organization is a continuous coproducer of anxiety and insecurity among those members who identify with the future. But members identify themselves with the past as well. They can see the image of their own future in the fate of those who in the past had been effective members of the system and served it well. An undesirable and unfortunate image is a serious source of insecurity that is at the core of alienation, corruption, and lust for power. This is why concern for the interests of past members, minimally in the form of an acceptable retirement system, is essential. In this respect the notion of gradual retirement, with all of its ramifications, should be considered more seriously than it has been.

Finally, the requirement for diachronic compatibility is based on the notion that a social system, unlike a biological system, is not necessarily subject to life cycles. The purpose of an organization is to serve its members by effectively serving its environment. This is a continuous and never ending function. Unlike a product-based organization, which will necessarily experience a life cycle, a learning and adaptive social system continuously recreates itself to meet the requirements of an everchanging environment and the changing desires of its members.

Performance Criteria, Measure of Performance, and Incentives

Thus far we have discussed integration in social systems through creation of vertical, horizontal, and diachronic compatibility of performance criteria. These criteria,

PERFORMANCE CRITERIA AS A MEANS OF SOCIAL INTEGRATION

however, are always operationalized by *measures* of performance and reinforced by the *incentive* system. In this section, we will elaborate on the distinction among these and discuss the role that each plays in the behavior of the system.

Performance criteria are expressions of the value system governing the organization. They establish the framework for the evaluation of the performance of the system. With these criteria, a set of relevant variables can be identified by which success of the system is evaluated.

Meaures of performance are the link between this framework and the concrete experience of evaluation and reward. These are the indicators by which the members are directly affected. They should be measured along the set of variables identified by the criteria, but this is seldom the case as we shall see.

The incentive system is a priority scheme that is superimposed on the measures of performance. It emphasizes certain measures and discounts others in order to produce the desired behavior. Incentive systems are often equated with explicit bonus programs and financial rewards. In reality, these are but a small part of such systems. The incentive system comprises all the various evaluations and inducements that affect the behavior of the members of the system: financial and nonfinancial; implicit or explicit, imposed by superiors, peers, and subordinates; and so on. Furthermore, the totality of this intricate system is seldom consciously designed. In fact parts of it that are designed, such as bonus programs, do not usually lead to expected outcomes because other intervening factors that can lead to counterintuitive behaviors are disregarded.

In practice the performance criteria are usually left vague, ambiguous, and unoperationalized. Even when their underlying concepts are relatively simple, such as minimization of costs in an organizational unit, they create complicated problems of measurement (for example, how to allocate the overhead). In the interest of uniformity of measures across the various systems, in both organizational and national contexts, the traditional approach has been to settle for simple, intuitive, and conventional measures.

The overriding concern has been with the accuracy of measures rather than with their relevance. Thus, the measures of performance are no longer justified by their relationship to the performance criteria. Instead, their justification is based on their simplicity, universal applicablity, and conventionality.

What seemed to be an innocent matter of convenience has changed the nature of measures of performance from navigational instruments to ends and objectives in and of themselves. Since we cannot measure what we want, we pretend to want what we can measure.

The irony of the situation is that the more accurate the measurement of the wrong performance criteria, the faster the road to disaster. The right criteria inaccurately measured, is much safer than the wrong criteria accurately measured. Let us explore the practical implications of this argument by means of two examples at the corporate and national levels.

In the case of a corporation, the usual practice of allocating overhead expenses to various operating units provides examples of unintended consequences of simple measures of performance. Despite the fact that overhead constitutes a major portion of the costs of the operating units (often running as high as one third of total costs), the criteria for allocation of this overhead are usually selected based on simplicity of measurement. Single factors such as the space occupied by a unit or the direct labor content of production, are among the most popular allocation criteria. Therefore, these variables naturally become the driving force for improvement of measures of performance. The fact that the allocation rule was simply an accounting convenience and that these measures are at best *estimates* of a fair share of common expenses, is easily forgotten. Once the allocation criteria becomes a rule, its relationship to generation of costs is by default assumed to be causal, as demonstrated by the following case.

Concerned with its cost disadvantage, a multinational manufacturing concern instructed its operating units to reduce their costs by 20%. The overhead costs in this company were allocated based on direct labor content of production. The rule was simply that overhead was equal to 300% of direct labor. The cost structure of the operating units were: Raw material and energy 40%, Direct labor 15%, Overhead 45%. Although direct labor contributed the lowest proportion of cost to the production process, it was targeted for reduction — because of the allocation rule of overhead costs. The operating units' managers realized that a 5% reduction in their direct labor costs would achieve the 20% "savings" required (it would reduce the allocated overhead by an additional 15%). The only problem was that when they all chose this path the corporation suddenly found itself with unallocated overhead. Although the operating units showed the 20% reduction, the corporation had not realized more than 5% real reduction of costs. The immediate reaction was the conventional one. The rate of overhead allocation was increased from 300% to 450%! Only as a result of the outrage that this action produced (and the impossible implications that it had of having to further reduce the labor costs) did the corporation finally decide to examine its overhead costs. The number of employees at the headquarters was subsequently reduced from 850 to 130.

This situation is by no means atypical. Because of the prevelance of allocation rules based on labor, pressure is unduly shifted to direct labor and the "knee jerk" reaction is to lay off productive manpower. If the police department is facing deficits, policemen are the first to be fired. If schools are in financial trouble, the number of teachers is reduced. Reduction of operating units does not automatically reduce overhead, as management seems to assume. On the contrary, it will increase the burden of

PERFORMANCE CRITERIA AS A MEANS OF SOCIAL INTEGRATION

the remaining units until the whole system comes to a halt.

Recently a large supermarket chain decided to close down ten of its stores because the accounting system showed that they were not covering the allocated overhead. Since the overhead did not decline proportionately, and in the case in point not at all, the other stores had to offset a larger share of overhead, which then made more of them unprofitable. Therefore, the company gradually withdrew more stores from the market, blaming high labor costs for its decision. When the stores reopened after some unprecedented labor initiatives with a new contract emphasizing participation of labor in decision making and establishment of a new subsidiary of the original corporation, there was no overhead allocated to the stores. Each store became responsible for its own operating budget, the balance of which was passed on to the corporation. The corporation, in turn, had to provide its services within the bounds of these surpluses. The extent of its services and the size of its bureaucracy thus reflected the needs of its operating units.

Perhaps the most dramatic instance of implications of measures of performance is a recent experience at a chemical company in which overhead was allocated based on the square footage of the facilities. A talented plant manager devised a new production process in his plant that eliminated the need for four large buildings. The machinery were moved out and the buildings vacated for other uses. Soon after, the manager realized there was no reduction in his overhead charges. The accounting department explained to him that overhead was allocated based on the space available to the plant; it did not matter whether the buildings were used or not. The next day bulldozers were tearing down the buildings! The overhead charges and total costs of the plant were reduced accordingly. In the meantime the company had lost perhaps millions of valuable assets.

A similar situation exists at the national level. Development is widely regarded as an objective of the social system. Initial struggle with designing measures for this complex process led to the conclusion that it would be difficult to measure development directly. It was noticed, however, that there are certain correlates of development that could be easily measured. Rate of growth of per capita income became the conventional measure of development. This identification grew so strong in the following years that the two concepts became identical in the minds of most people. Societies around the world became pre-occupied with increasing their per capita income. There was no longer a need to be concerned with development because of this simple substitution.

In the early seventies an increase in oil prices produced instantly developed nations, according to the accepted measures. The shock that this prouded resulted in some rethinking about the measures of development. However, the concepts of correlates was so embedded in the concerned community that it was not questioned. The solution was merely to increase the number of correlates being measured. We now have a series of these indicators that substitute for development, such as per capita steel production, number of doctors per 1000 population, per capita consumption of fuel. It is not surprising, then, to find national devlopment policies aimed at improving these measures, usually at incredible cost to the society at large.

Performance Center

Since the members of an organization are purposeful systems, their patterns of interaction cannot be predefined and therefore, are not subject to a mechanistic concept of design. Actors (individually or in groups) by agreeing or disagreeing with each other on the compatibility of their ends, means, or both may at the same time cooperate with regard to one set of objectives, compete over another, and be in conflict with respect to a third.

In the systems view, performance of the whole is not the sum of the performances of its parts, rather it is the product of their interaction. The proper management of this interaction defines the main role of a manager, a role that cannot be replaced, mechanized, or predefined. This is why the ultimate success of an organization depends on its management talent. However, proliferation of variety produced by highly differentiated functional structures presents a severe challenge to this critical management function. So much so that the law of diminishing returns is increasingly overtaking the advantages of economies of scale. In this context an organizational culture that promotes the cooperative type of interactions is the most important help a manager can have.

By rejecting unquestionable acceptance of zero-sum-situations, the systems view of organization considers both lose/lose and win/win situations to be strong possibilities, therefore the objective is to create an organizational setting that will improve the chances of win/win situations.

Among the fundamental decisions to be made in the design of an organization is how authority and responsibility will be allocated. Any large organization contains smaller units each of which has a defined responsibility and a certain level of authority.

More difficult than allocating authority and responsibility is building these into the design in proper combination. Too often units that have the authority to make decisions in organizations have no responsibility for the consequences of those decisions. Conversely, units may be held responsible for situations over which they have no control. Both situations result in frustration and ineffectiveness.

PERFORMANCE CRITERIA AS A MEANS OF SOCIAL INTEGRATION

A U T H O R I T Y	high authority low responsibility	high authority high responsibility
	low authority low reponsibility	low authority high responsibility

RESPONSIBILITY

In organizations where the high-high quadrant is unknown or rare, the name of the game becomes avoidance of the low authority-high responsibility quadrant and jockeying for the high authority-low responsibility positions. These positions soon become competitive and relatively unattainable. Thus, most members of the organizations drift toward the low-low quadrant in preference to the only alternative, low authority-high responsibility. This results in an organizational culture of "just following orders" and "passing the buck;" a bureaucratic culture with ultimate lose/lose results.

Traditionally, control in organizations is achieved through a hierarchical reporting relationship. In this approach, optimization of performance of the higher-level unit requires that the lower level's objectives be subordinated. For example, if a marketing unit requires low manufacturing costs, it can best assure this by direct control over manufacturing.

Control, however, can be achieved in at least one other way; an exchange or interactive relationship. The model for this is the relationship between suppliers and consumers. Although consumers hold no authority to direct the supplier to produce a specific product, they exert considerable influence on the decision by placing orders for that product. Units within an organization can also relate to each other in an exchange relationship. However, for an exchange relationship to work, a vertical, horizontal, and diachronic compatibility between the performance criteria of different units must be assured. These, along with relevant performance measures and an appropriate incentive system are required for interactions among units to result in win/win situations. As argued earlier, a pure profit center, although useful for providing horizontal compatibility, has serious shortcomings with regard to diachronic compatibility. There is the potential for profit centers to look to short-term results, to focus on the single objective of profit, and to disregard opportunities that might have short-term negative consequences for the center, but which could provide significant long-term gains. The following concept of performance centers is intended to overcome these deficiencies.

Effectiveness of a performance center is reflected not only by its efficiency (input, output ratios), but also by its latency (the future potential).

	Low	High
High	L, H	H, L
Low	L, L	H, L

Efficiency (rows) / Latency (columns)

It is obvious that current performance of a system can be improved at the expense of its future (L,H), and vice versa (H,L). But it is not so obvious that a system can be a low performer today and be obstructing its future as well (L,L). More important, not only is there a possibility for a High, High situation, but it is a necessity.

The critical point of this conception is that the market value of a performance center, that is, the present value of its future earnings together with its present profit and loss performance, becomes an important measure of its performance. By accepting long-term responsibility for its viability, a performance center, will become sensitive to the marketing consequences of its actions. Thus, a market orientation is developed and internalized. The necessary change of emphasis from products to markets will reduce the system's dependency on product life cycle and create an entrepreneurial culture, which will insure its continuity and enhance its diachronic compatibility.

A performance center, like a profit center, has profit and loss responsibility. Thus, its concern is with adding value not minimizing cost or maximizing revenue. However, critical to the successful operation of a performance center is its ability to have partial control or influence over determining its (1) product mix, (2) pricing of outputs, (3) selection of sources of inputs, and (4) ability to retain part of its earning for future development.

Finally, an incentive system is needed to encourage the search for win/win solutions by creating positive value to members as a result of cooperative strategies. An organizational unit or member should be rewarded not only for improvement in measure of performance, but also for any contribution made to other units that improves their measures. An example of this is provided by the following scheme implemented in a professional firm.

All members received 10% of their annually generated income if certain performance measures were met. This percentage, however, would be increased to 15% if other members of a group would achieve their objectives and would increase to 20% if all the various groups were able to meet their performance objectives.

PERFORMANCE CRITERIA AS A MEANS OF SOCIAL INTEGRATION

Unfortunately, all this is easier said than done. The majority of present corporate citizens are too happy with their perceived sense of security, embedded in the low authority and low responsibility culture of bureaucratic corporations, to permit any meaningful change in the system without the active support and commitment of its top level management. However, our previous experiences show that the concept can more easily be introduced in the formative stages of new corporations.

To summarize, we propose that the incompatibility of performance criteria among the units within an organization is the main source of the frustration and dilemmas encountered in most organizational settings. Proliferation of these incompatibilities, at least partially, explains the reason why the famous "invisible hand" has been converted to an "invisible kick."

Reflection on the nature and working mechanism of the "invisible hand" would reveal that it is the vertical, horizontal, and diachronic compatibility of performance criteria. Note that implicit in the notion of the invisible hand is a uniform performance criteria. Adam Smith assumed that all actors are profit centers. To conclude, the answer to the question, Why is the system in such a mess now when it worked so well before? is provided by Boudling's famous assertion, "The name of the devil is suboptimization."[4]

REFERENCE

1. Ackoff, R. L. and Emery, F.E., *On Purposeful Systems*, Intersystems Publications, Seaside, 1981.

2. Ackoff, R. L., Finnel, E. V. and Gharajedaghi, J., *A Guide to Controlling Your Corporation's Future*, John Wiley & Sons, New York, 1984.

3. Beer, Stafford, *Brain of the Firm*, Penguin Press, 1972.

4. Boulding, K., *Beyond Economics*, The University of Michigan Press, Ann Arbor.

5. Gharajedaghi, J. and Ackoff, R., "Mechanisms, Organisms and Social Systems," *Strategic Management Journal, Vol. 5*. John Wiley & Sons, New York, 1984.

BIBLIOGRAPHY

1. Ackoff, Russell, *Redesigning the Future*. New York: John Wiley & Sons, 1974.
2. Ackoff, Russell, *Art of Problem Solving*. New York: John Wiley & Sons, 1978.
3. Ackoff, Russell & Emery, Fred E., *On Purposeful Systems*. Seaside: Intersystems Publications, 1981.
4. Althusser, L., *For Marx*. Harmondsworth: Penguin Press, 1969.
5. Althusser, L. & Balibar, E.,. *Reading Capital*. London: New Left Books, 1970.
6. Argyris, C. & Schon, D. A., *Theory in Practice*. San Francisco: Jossey-Bass Publishers, 1975.
7. Ashby, W. Ross, *An Introduction to Cybernetics*. Chapman and Hall, 1956.
8. Ashby, W. Ross, "General Systems Theory as a New Discipline," *General Systems 3: 1-6*, 1958.
9. Beer, Stafford, *Brain of the Firm*. Harmondsworth: Penguin Press, 1967.
10. Beer, Stafford, *Platforms of Change*. New York: John Wiley & Sons, 1979.
11. Beishon, John & Peters, Geoff (eds), *Systems Behaviour*. London: The Open University Press, 1972.
12. Bernstein, Henry (ed), *Underdevelopment & Development: The Third World Today*. Middlesex: Penguin Books, 1973.
13. Bertalanffy, Ludwig Von, *Robots, Man and Minds*. New York: Braziller, 1967.
14. Bertalanffy, Ludwig Von, *General Systems Theory: Foundation, Development, Applications*. Middlesex: Penguin Books, 1968.
15. Blake, R. R. & Mouton, J. S., *The Managerial Grid*. Houston: Gulf Publishing Company, 1964.
16. Bogdanov, A., *Essays in Tektology*, Translated by George Gorelik. Seaside: Intersystems Publications, 1980.
17. Boulding, Kenneth E., *The Image*. Ann Arbor: University of Michigan Press, 1956a.
18. Boulding, Kenneth E., *Ecodynamics*. Beverly Hills: Sage Publications, 1981.
19. Boulding, Kenneth E., *Beyond Economics*. Ann Arbor: The University of Michigan Press, 1968.
20. Bowler, T. Downing, *General Systems Thinking*. Holland/New York: Elsevier North, 1981.
21. Buckley, Walter, *Sociology and Modern Systems Theory*. Englewood Cliffs: Prentice-Hall, 1967.

22. Buckley, Walter (ed), *Systems Research for the Behavioral Scientist: A Source-Book*. Chicago: Aldine Publishing Co., 1968.
23. Churchman, C. West, *The Systems Approach and Its Enemies*. New York: Basic Books, 1979.
24. Churchman, C. West, *The Systems Approach*. New York: Delacorte Press, 1968.
25. Colletti, L., *From Rousseau to Lenin*. London: New Left Books, 1972.
26. Comte, Auguste, *The Positivist Philosophy, Vol. 1* (freely translated by H. Martineau). London: Chapman, 1853.
27. Dahrendorf, R., *Class and Class Conflict in Industrial Society*. London: Routledge and Kegan Paul, 1959.
28. Durkheim, E., *The Rules of Sociological Method*. Glencoe: Free Press, 1938.
29. Easton, David, *A Framework for Political Analysis*. Englewood Cliffs: Prentice-Hall, 1965a.
30. Easton, David, *A Systems Analysis of Political Life*. New York: University of Chicago/Free Press, 1965b.
31. Eisenstadt, S. N., *Modernization: Protest and Change*. Englewood Cliffs: Prentice-Hall, 1966.
32. Emery, F. E., *Systems Thinking*. Middlesex: Penguin Books, 1969.
33. Emery, F. E. and Trist, E. L., *Towards a Social Ecology*. Harmondsworth: Penguin Press, 1972.
34. Fayol, H., *General and Industrial Management* (translated by C. Storr). London: Pitman, 1949.
35. Forrester, Jay W., *Industrial Dynamics*. Cambridge: Wright-Allen Press, 1961.
36. Forrester, Jay W., *World Dynamics*. Cambridge: Wright-Allen Press, 1971.
37. Freud, Sigmund, *Civilization and Its Discontents*. Garden City: Doubleday, 1930.
38. Fromm, Eric, *The Sane Society*. New York: Rinehart, 1955.
39. Gouldner, A. W., *The Dialectic of Ideology and Technology*. London and New York: Macmillan, 1976.
40. Habermas, J., *Towards a Rational Society*. London: Heinemann, 1971a.
41. Habermas, J., *Knowledge and Human Interests*. London: Heinemann, 1972.
42. Habermas, J., *Theory and Practice*. London: Heinemann, 1974.
43. Habermas, J., *Legitimation Crisis*. London: Heinemann, 1976.
44. Hall, A. D. & Fagen, R. E., "Definition of System," *General Systems*, 1959.
45. Hegel, G., *The Phenomenology of Mind*. London: George Allen and Unwin, 1931.
46. Homans, G. C., *Social Behaviour: Its Elementary Forms*. New York: Harcourt Brace and World, 1961.
47. Kahn, R. L. and Boulding, E. (eds), *Power and Conflict in Organisations*. London: Tavistock Publications, 1964.
48. Katz, D. and Kahn, R. L., *The Social Psychology of Organisations*. New York: John Wiley, 1966.
49. Kaplan, Abraham, *The Conduct of Inquiry*. New York: Harper & Row, 1964.
50. Keynes, John Maynard, *General Theory of Employment, Interest and Money*. New York: Martin Press, 1936.

BIBLIOGRAPHY

51. Laszlo, Ervin, *The Systems View of the World*. New York: George Braziller, 1972a.
52. Laszlo, Ervin, *Introduction to Systems Philosophy: Toward a New Paradigm of Contemporary Thought*. New York: George Braziller, 1972b.
53. McGregor, D., *The Human Side of Enterprise*. New York: McGraw Hill, 1960.
54. McClelland, David, *The Achieving Society*. New York: Free Press, 1961.
55. March, J. G., *Handbook of Organizations*. Chicago: Rand McNally, 1965.
56. March, J. G. and Simon, H. A., *Organisations*. New York: John Wiley, 1958.
57. Marcuse, H., *One-Dimensional Man*. London: Routledge and Kegan Paul, 1964.
58. Meadows, Dennis L. & Meadows, Donella H., *Toward Global Equilibrium: Collected Papers*. Cambridge: Wright-Allen Press.
59. Meadows, Donella H., et al, *The Limits to Growth*. London: Potomac Associates, 1972.
60. Mill, John Stuart, *Principles of Political Economy*. London: Parker, 1848.
61. Miller, James G., "Toward a General Theory for the Behavioral Sciences," *American Psychologist, 10*: 513-531, 1955.
62. Miller, James, *Living Systems*. New York: McGraw Hill, 1978.
63. Parsons, T., *The Social System*. London: Tavistock; Glencoe: Free Press.
64. Piaget, J., *Structuralism*. New York: Harper, Torch Books, 1971.
65. Rapoport, Anatol & Horvath, William J., "Thoughts on Organization Theory," *General Systems, 4*, 1959, 87-91.
66. Rapoport, Anatol, "General Systems Theory," *The International Encyclopaedia of Social Sciences*, edited by David L. Sills. New York: The Macmillan Co. & The Free Press, 1968.
67. Rex, J., *Approaches to Sociology*. London: Routledge and Kegan Paul, 1974.
68. Ricardo, David, *Principles of Political Economy*. London: J. Murray, 1817.
69. Rostow, W. W., *The Stages of Economic Growth: A Non-Communist Manifesto*. Cambridge: Cambridge University Press.
70. Sartre, J. P., *Between Existentialism and Marxism*. London: Pantheon, 1974.
71. Sartre, J. P., *Critique of Dialectical Reasons, vol. I*. London: New Left Books, 1976.
72. Schon, D., *Beyond Stable State*. New York: Basic Books, 1971.
73. Schumpeter, Joseph A., *History of Economic Analysis*. New York: Oxford University Press, 1954.
74. Simon, H. A., *Administrative Behaviour: A Study of Decision Making Processes in Administrative Organisation* (2nd ed). New York: Collier/Macmillan, 1957.
75. Simon, H., "On the Concept of Organisational Goal," *Administrative Science Quarterly, 9 (1)*. 1964, pp. 1-22.
76. Skinner, B. F., *Beyond Freedom and Dignity*. New York: Alfred Knopf, 1972.
77. Smith, Adam, *An Inquiry into the Nature and Causes of the Wealth of Nations*. New York: Modern Library, 1937.
78. Spencer, Herbert, *The Principles of Sociology*, 3 vols. New York: Appleton, (1925-29).
79. Thurow, Lester, C., *Zero Sum Society*. New York: Basic Books, 1980.
80. Vickers, G., *The Art of Judgement*. London: Chapman and Hall, 1966.

81. Vickers, G., *Value Systems and Social Processes*. Middlesex: Penguin Books, 1968.
82. Weber, M., *The Theory of Social and Economic Organisation* (translated by A. Henderson and T. Parsons). Glencoe: Free Press, 1947.
83. Weber, M., *The Methodology of the Social Sciences*. Glencoe: Free Press, 1949.
84. Wiener, N., *The Human Use of Human Beings*. New York: Doubleday Anchor, 1954.
85. Zeeman, E.C., "Catastrophe Theory," *Scientific American, 234 (4)*, 1976, pp. 65-83.
86. Zeleny, M. (ed), *Autopoiesis*. Amsterdam: North Holland Publishing Company, 1981.